CAMBRIDGE LIBRARY COLLECTION

Books of enduring scholarly value

Physical Sciences

From ancient times, humans have tried to understand the workings of the world around them. The roots of modern physical science go back to the very earliest mechanical devices such as levers and rollers, the mixing of paints and dyes, and the importance of the heavenly bodies in early religious observance and navigation. The physical sciences as we know them today began to emerge as independent academic subjects during the early modern period, in the work of Newton and other 'natural philosophers', and numerous sub-disciplines developed during the centuries that followed. This part of the Cambridge Library Collection is devoted to landmark publications in this area which will be of interest to historians of science concerned with individual scientists, particular discoveries, and advances in scientific method, or with the establishment and development of scientific institutions around the world.

Memoirs of Dr. Joseph Priestley, to the Year 1795

Joseph Priestley (1733–1804) was an eighteenth-century English polymath with accomplishments in the fields of science, pedagogy, philosophy and theology. Among his more notable achievements were the discovery of oxygen and his work in establishing Unitarianism. Often a controversialist, Priestley's efforts to develop a 'rational' Christianity and support for the French Revolution eventually made him unwelcome in his native land. His 1807 *Memoirs* relate the story of his life until the time of his 1794 emigration to America and include other biographical materials written by his son. This first volume also contains five appendices discussing his philosophy, scientific work and religious opinions. Priestley's memoirs are an important source for anyone interested in the state of epistemology, rationalism, and religious belief in the age of the Enlightenment, and in a man who, in the words of his son, 'gave unremitting exertions in the cause of truth'.

Cambridge University Press has long been a pioneer in the reissuing of out-of-print titles from its own backlist, producing digital reprints of books that are still sought after by scholars and students but could not be reprinted economically using traditional technology. The Cambridge Library Collection extends this activity to a wider range of books which are still of importance to researchers and professionals, either for the source material they contain, or as landmarks in the history of their academic discipline.

Drawing from the world-renowned collections in the Cambridge University Library, and guided by the advice of experts in each subject area, Cambridge University Press is using state-of-the-art scanning machines in its own Printing House to capture the content of each book selected for inclusion. The files are processed to give a consistently clear, crisp image, and the books finished to the high quality standard for which the Press is recognised around the world. The latest print-on-demand technology ensures that the books will remain available indefinitely, and that orders for single or multiple copies can quickly be supplied.

The Cambridge Library Collection will bring back to life books of enduring scholarly value (including out-of-copyright works originally issued by other publishers) across a wide range of disciplines in the humanities and social sciences and in science and technology.

Memoirs of Dr. Joseph Priestley, to the Year 1795

With a Continuation, to the Time of his Decease by his Son, Joseph Priestley

VOLUME 1

JOSEPH PRIESTLEY

CAMBRIDGE UNIVERSITY PRESS

Cambridge, New York, Melbourne, Madrid, Cape Town, Singapore,
São Paolo, Delhi, Dubai, Tokyo

Published in the United States of America by Cambridge University Press, New York

www.cambridge.org
Information on this title: www.cambridge.org/9781108014199

© in this compilation Cambridge University Press 2010

This edition first published 1806
This digitally printed version 2010

ISBN 978-1-108-01419-9 Paperback

MEMOIRS

OF

DR. JOSEPH PRIESTLEY,

TO THE YEAR 1795,

WRITTEN BY HIMSELF:

WITH A CONTINUATION, TO THE TIME OF HIS DECEASE,

BY HIS SON, JOSEPH PRIESTLEY:

AND OBSERVATIONS ON HIS WRITINGS,

BY THOMAS COOPER, PRESIDENT JUDGE OF THE

4TH DISTRICT OF PENNSYLVANIA: AND THE

REV. WILLIAM CHRISTIE.

LONDON:

PRINTED FOR J. JOHNSON, NO. 72, ST. PAUL'S CHURCH-YARD,

1806.

ADVERTISEMENT.

THE Analysis of my father's Theological writings mentioned in the Preface to this work, is in the press and will be printed in the same manner as the Memoirs now presented to the public, and may be purchased with the Catalogue of his writings separately to bind with the present Volume, or may be had bound up with four Sermons which my father desired me to print, making therewith a second volume.

I had an expectation of presenting the public with an Engraving of a striking likeness of my father, to be prefixed to the present volume. In this expectation I have been disappointed. I hope however to be able to do it, by the time the second Volume will make its appearance.

J. P.

1

TABLE OF CONTENTS.

PREFACE.

—◆—

$\mathbf{M}$Y father, Dr. Priestley, having taken the trouble of writing down the principal occurrences of his life, to the period of his arrival in this country, that account is now presented to the public in the state in which he left it, one or two trifling alterations excepted. The simple unaffected manner in which it is written, will be deemed, I have no doubt, far more interesting, than if the narrative itself had been made the text of a more laboured composition.

Independent of the desire, so universal among mankind, to know somewhat of the private as well as the public history of those who have made themselves eminent among their fellow citizens, the life of my father is likely to be more useful as well as more interesting than those of the generality of literary men ; not only as it is an account of great industry combined with great abilities, successfully exerted for the extension of human improvement, but because it affords a striking proof of the value

of

of rational christianity, adopted upon mature reflection and practiced with habitual perseverance.

Few men have had to struggle for so many years with circumstances more straitened and precarious than my father; few men have ventured to attack so many or such inveterate prejudices respecting the prevalent religion of his country, or have advanced bolder or more important opinions in opposition to the courtly politics of the powers that be; few have had to encounter more able opponents in his literary career, or have been exposed to such incessant and vindictive obloquy, from men of every description, in return for his unremitting exertions in the cause of truth; yet none have more uniformly proceeded with a single eye, regardless of consequences, to act as his conviction impelled him, and his conscience dictated. His conduct brought with it its own reward, reputation and respect from the most eminent of his contemporaries, the affectionate attachment of most valuable friends, and a cheerfulness of disposition arising in part from conscious rectitude which no misfortunes could long repress. But to me it seems, that conscious rectitude alone would hardly, of itself, have been able to support him under some

of

of the afflictions he was doomed to bear. He had a farther resource, to him never failing and invaluable, a firm persuasion of the benevolence of the Almighty towards all his creatures, and the conviction that every part of his own life, like every part of the whole system, was preordained for the best upon the whole of existence. Had he entertained the gloomy notions of Calvinism in which he was brought up, this cheering source of contentment and resignation would probably have failed him, and irritation and despondency would have gained an unhappy ascendancy. But by him the deity was not regarded as an avenging tyrant, punishing, for the sake of punishing his weak and imperfect creatures, but as a wise and kind parent, inflicting those corrections only that are necessary to bring our dispositions to the proper temper, and to fit us for the highest state of happiness of which our natures are ultimately capable.

With these views of the present and the future, it is no wonder that he submitted with perfect resignation to the inevitable vicissitudes of human life, and looked forward to futurity, as a period of existence when his capacity for receiving happiness would be greater because his capacity for communicating it would be enlarged.

My

My father's narrative closing with his arrival in
this country, where he has done so much for the promotion of useful knowledge of all kinds, I have compleated the account of his life from that period to the
termination of it. The Notes have been added to
the narrative as desireable illustrations of the passages to which they refer.

I have likewise thought it proper to add a review
of my father's literary labours, in order to give the
reader a knowledge of his opinions on many important subjects, likewise, of the share in the increase of
human knowledge, which may be justly ascribed to
his exertions. The Appendices giving an account of
his Chemical, Philosophical, Metaphysical, Political
and Miscellaneous writings, as well as the Summary
of his religious opinions, are written by my friend
Judge Cooper, formerly of Manchester in England.
For the Appendix containing an analysis of my father's Theological writings, I am indebted to the
Rev. W. Christie, formerly of Montrose in Scotland.

The work might have been made more interesting
as well as entertaining, had I deemed myself at liberty to have published letters addressed to my father

by

by persons of eminence in this country, as well as in Europe. But those communications that were intended to be private, shall remain so; as I do not think I have a right to amuse the public either against, or without, the inclinations of those who confided their correspondence to his care.

I regret, that more of the present work is not the production of my father's pen; and I hope the reader will make allowance for the imperfection of that portion of it, for which I have made myself responsible.

JOSEPH PRIESTLEY.

Northumberland, Pennsylvania,

May 1st, 1805.

MEMOIRS

OF

Dr. JOSEPH PRIESTLEY.

[WRITTEN BY HIMSELF.]

HAVING thought it right to leave behind me some account of my *friends* and *benefactors*, it is in a manner necessary that I also give some account of *myself*; and as the like has been done by many persons, and for reasons which posterity has approved, I make no farther apology for following their example. If my writings in general have been useful to my cotemporaries, I hope that this account of myself will not be without its use to those who may come after me, and especially in promoting virtue and piety, which I hope I may say it has been my care to practice myself, as it has been my business to inculcate them upon others.

A   My

My father, Jonas Priestley, was the youngest son of Joseph Priestley, a maker and dresser of woollen cloth. His first wife, my mother, was the only child of Joseph Swift, a farmer at Shafton, a village about six miles south east of Wakefield. By this wife he had six children, four sons and two daughters. I, the oldest, was born on the thirteenth of March, old style 1733, at Fieldhead about six miles south west of Leeds in Yorkshire. My mother dying in in 1740, my father married again in 1745, and by his second wife had three daughters.

My mother having children so fast, I was very soon committed to the care of her father, and with him I continued with little interruption till my mother's death,

It is but little that I can recollect of my mother. I remember, however, that she was careful to teach me the Assembly's Catechism, and to give me the best instructions the little time that I was at home. Once in particular, when I was playing with a pin, she asked me where I got it; and on telling her that I found it at my uncle's, who lived very near to my father, and where I had been playing with my cousins, she made me carry it back again; no

doubt

doubt to impress my mind, as it could not fail to do, with a clear idea of the distinction of property, and of the importance of attending to it. She died in the hard winter of 1739, not long after being delivered of my youngest brother ; and having dreamed a little before her death that she was in a delightful place, which she particularly described, and imagined to be heaven, the last words she spake, as my aunt informed me, were " Let me go to that fine " place."

On the death of my mother I was taken home, my brothers taking my place, and was sent to school in the neighbourhood. But being without a mother, and my father incumbered with a large family, a sister of my fathers, in the year 1742, relieved him of all care of me, by taking me entirely to herself, and considering me as her child, having none of her own. From this time she was truly a parent to me till her death in 1764.

My aunt was married to a Mr. Keighly, a man who had distinguished himself for his zeal for religion and for his public spirit. He was also a man of considerable property, and dying soon after I went to them, left the greatest part of his fortune to my aunt for

life,

life, and much of it at her disposal after her death.

By this truly pious and excellent woman, who knew no other use of wealth, or of talents of any kind, than to do good, and who never spared herself for this purpose, I was sent to several schools in the neighbourhood, especially to a large free school, under the care of a clergyman, Mr. Hague, under whom, at the age of twelve or fifteen, I first began to make any progress in the Latin Tongue, and acquired the elements of Greek. But about the same time that I began to learn Greek at this public school, I learned Hebrew on holidays of the dissenting minister of the place, Mr Kirkby, and upon the removal of Mr. Hague from the free school, Mr. Kirkby opening a school of his own, I was wholly under his care. With this instruction I had acquired a pretty good knowledge of the learned languages at the age of sixteen. But from this time Mr. Kirkby's increasing infirmities obliged him to relinquish his school, and beginning to be of a weakly consumptive habit, so that it was not thought adviseable to send me to any other place of education, I was left to conduct my studies as well as I could

till

till I went to the academy at Daventry in the year 1752.

From the time I discovered any fondness for books my aunt entertained hopes of my being a minister, and I readily entered into her views. But my ill health obliged me to turn my thoughts another way, and with a view to trade, I learned the modern languages, French, Italian, and High Dutch without a master ; and in the first and last of them I translated, and wrote letters, for an uncle of mine who was a merchant, and who intended to put me into a counting house in Lisbon. A house was actually engaged to receive me there, and every thing was nearly ready for my undertaking the voyage. But getting better health my former destination for the ministry was resumed, and I was sent to Daventry, to study under Mr. Ashworth, afterwards Dr. Ashworth.

Looking back, as I often do, upon this period of my life, I see the greatest reason to be thankful to God for the pious care of my parents and friends, in giving me religious instruction. My mother was a woman of exemplary piety, and my father also had a strong sense of religion, praying with his family morning and evening, and carefully teaching his chil-

dren

dren and servants the Assembly's Catechism, which was all the system of which he had any knowledge. In the latter part of his life he became very fond of Mr. Whitfield's writings, and other works of a similar kind, having been brought up in the principles of Calvinism, and adopting them, but without ever giving much attention to matters of speculation, and entertaining no bigotted aversion to those who differed from him on the subject.

The same was the case with my excellent aunt, she was truly Calvinistic in principle, but was far from confining salvation to those who thought as she did on religious subjects. Being left in good circumstances, her home was the resort of all the dissenting ministers in the neighbourhood without distinction, and those who were the most obnoxious on account of their heresy were almost as welcome to her, if she thought them honest and good men, (which she was not unwilling to do) as any others.

The most heretical ministers in the neighbourhood were Mr. Graham of Halifax, and Mr. Walker of Leeds, but they were frequently my Aunt's guests. With the former of these my intimacy grew with my years, but chiefly after I became a preacher. We
kept

kept up a correspondence to the last, thinking alike on most subjects. To him I dedicated my *Disquisitions on Matter and Spirit*, and when he died, he left me his manuscripts, his Polyglot bible, and two hundred pounds. Besides being a rational christian, he was an excellent classical scholar, and wrote Latin with great facility and elegance. He frequently wrote to me in that language.

Thus I was brought up with sentiments of piety, but without bigotry, and having from my earliest years given much attention to the subject of religion, I was as much confirmed as I well could be in the principles of Calvinism, all the books that came in my way having that tendency.

The weakness of my constitution, which often led me to think that I should not be long lived, contributed to give my mind a still more serious turn, and having read many books of *experiences*, and in consequence believing that a *new birth* produced by the immediate agency of the Spirit of God, was necessary to salvation, and not being able to satisfy myself that I *had* experienced any thing of the kind, I felt occasionally such distress of mind as it is not in my power to describe, and which I still look back

upon with horror. Notwithstanding I had nothing
very material to reproach myself with, I often con-
cluded that God had forsaken me, and that mine was
like the case of Francis Spira, to whom, as he imagined, repentance and salvation were denied. In
that state of mind I remember reading the account
of the man in the iron cage in the Pilgrim's Progress
with the greatest perturbation.

I imagine that even these conflicts of mind were
not without their use, as they led me to think habitually of God and a future state. And though my
feelings were then, no doubt, too full of terror,
what remained of them was a deep reverence for divine things, and in time a pleasing satisfaction
which can never be effaced, and I hope, was strength-
ened as I have advanced in life, and acquired more
rational notions of religion. The remembrance,
however, of what I sometimes felt in that state of
ignorance and darkness gives me a peculiar sense of
the value of rational principles of religion, and of
which I can give but an imperfect description to
others.

As *truth*, we cannot doubt, must have an advantage over *error*, we may conclude that the want of
these

these peculiar feelings is compensated by something of greater value, which arises to others from always having seen things in a just and pleasing light; from having always considered the Supreme Being as the kind parent of all his offspring. This, however, not having been my case, I cannot be so good a judge of the effects of it. At all events, we ought always to inculcate just views of things, assuring ourselves that *proper feelings and right conduct* will be the consequence of them.

In the latter part of the interval between my leaving the grammar school and going to the academy, which was something more than two years, I attended two days in the week upon Mr. Haggerstone, a dissenting minister in the neighbourhood, who had been educated under Mr. Maclaurin. Of him I learned Geometry, Algebra and various branches of Mathematics, theoretical and practical. And at the same time I read, but with little assistance from him, Gravesend's Elements of Natural Philosophy, Watt's Logic, Locke's Essay on the Human Understanding, &c, and made such a proficiency in other branches of learning, that when I was admitted at the academy (which was on Coward's foundation) I was excused

cused all the studies of the first year, and a great part of those of the second.

In the same interval I spent the latter part of every week with Mr. Thomas, a baptist minister now of Bristol but then of Gildersome, a village about four miles from Leeds, who had had no learned education. Him I instructed in Hebrew, and by that means made myself a considerable proficient in that language. At the same time I learned Chaldee and Syriac, and just began to read Arabic. Upon the whole, going to the academy later than is usual, and being thereby better furnished, I was qualified to appear there with greater advantage.

Before I went from home I was very desirous of being admitted a communicant in the congregation which I had always attended, and the old minister, as well as my Aunt, were as desirous of it as myself, but the elders of the Church, who had the government of it, refused me, because, when they interrogated me on the subject of the *sin of Adam*, I appeared not to be quite orthodox, not thinking that all the human race (supposing them not to have any sin of their own) were liable to the wrath of God, and the pains of hell for ever, on account of
that

that sin only ; for such was the question that was put to me.   Some time before, having then no doubt of the truth of the doctrine, I well remember being much distressed that I could not feel a proper repentance for the sin of Adam ; taking it for granted that without *this* it could not be forgiven me.   Mr. Haggerstone above mentioned, was a little more liberal than the members of the congregation in which I was brought up, being what is called a *Baxterian* ;*

and

* BAXTERIANS, The famous Non-conformist Richard Baxter who flourished about the middle of the last Century, attempted a Coalition between the doctrines of Calvin and Arminius. The former of these held that God from the beginning had elected a few of the human race to be saved, without reference to their good actions in this life, and had left the rest of mankind in a state of final and inevitable reprobation.   The latter was of opinion that the Christian dispensation furnished the means of final Salvation to all men, though the merits of the death of Christ would be ultimately advantageous to believers only. Baxter, thought with Calvin that some among mankind were from the beginning elected unto eternal life, and gifted from above with the saving grace necessary in the first instance to the several steps of a believer's christian character ; but he thought also with Arminius that all men had common grace imparted to them, sufficient to enable them if they chose, to attain unto final Salvation by using the means ordained by Christ and his Apostles.   Calvin also held the fi-

nal

and his general conversation had a liberal turn, and such as tended to undermine my prejudices. But what contributed to open my eyes still more was the conversation of a Mr. Walker, from Ashton under line, who preached as a candidate when our old minister was superannuated. He was an avowed Baxterian, and being rejected on that account his opinions were much canvassed, and he being a guest at the house of my Aunt, we soon became very intimate, and I thought I saw much of reason in his sentiments. Thinking farther on these subjects, I was, before I went to the academy, an Arminian, but had by no means rejected the doctrine of the trinity, or that of atonement.

Though after I saw reason to change my opinions I found

---

nal perseverance of the Saints, or as it has since been expressed that a believer might fall foully but not finally, whereas Baxter seems to have thought that not every one who had saving grace imparted to him would persevere to the end, or as the Arminian Methodists quaintly express it, he held that a believer may fall both foully and finally. The compromising doctrine of Baxter may be seen in his very learned and unintelligible work entitled Catholick Theology. He used to be an annual communicant in the Church of England by way of exemplifying his accommodating opinions. T. C.

I found myself incommoded by the rigour of the congregation with which I was connected, I shall always acknowledge with great gratitude that I owe much to it. The business of religion was effectually attended to in it. We were all catechized in public 'till we were grown up, servants as well as others: the minister always expounded the scriptures with as much regularity as he preached, and there was hardly a day in the week, in which there was not some meeting of one or other part of the congregation, On one evening there was a meeting of the young men for conversation and prayer. This I constantly attended, praying extempore with others when called upon.

At my Aunt's there was a monthly meeting of wo men, who acquitted themselves in prayer as well as any of the men belonging to the congregation. Being at first a child in the family, I was permitted to attend their meetings, and growing up insensibly, heard them after I was capable of judging. My Aunt after the death of her husband, prayed every morning and evening in her family, until I was about seventeen, when that duty devolved upon me.

The Lord's day was kept with peculiar strictness.

No

No victuals were dressed on that day in any family. No member of it was permitted to walk out for recreation, but the whole of the day was spent at the public meeting, or at home in reading, meditation, and prayer, in the family or the closet.

It was my custom at that time to recollect as much as I could of the sermons I heard, and to commit it to writing. This practice I began very early, and continued it until I was able from the heads of a discourse to supply the rest myself. For not troubling myself to commit to memory much of the amplification, and writing at home almost as much as I had heard, I insensibly acquired a habit of composing with great readiness; and from this practice I believe I have derived great advantage through life; composition seldom employing so much time as would be necessary to write in long hand any thing I have published.

By these means, not being disgusted with these strict forms of religion as many persons of better health and spirits probably might have been (and on which account I am far from recommending the same strictness to others) I acquired in early life a serious turn of mind. Among other things I had at

this

this time a great aversion to *Plays and Romances*, so that I never read any works of this kind except Robinson Crusoe, until I went to the academy. I well remember seeing my brother Timothy reading a book of Knight Errantry, and with great indignation I snatched it out of his hands, and threw it away. This brother afterwards, when he had for some time followed my fathers business (which was that of a Cloth-dresser) became, if possible, more serious than I had been; and after an imperfect education, took up the profession of a minister among the Independents, in which he now continues.

While I was at the Grammar School I learned *Mr. Annet's Short hand*, and thinking I could suggest some improvements in it, I wrote to the Author, and this was the beginning of a correspondence which lasted several years. He was, as I ever perceived, an unbeliever in Christianity and a necessarian. On this subject several letters, written with care on both sides, passed between us, and these Mr. Annet often pressed me to give him leave to publish, but I constantly refused. I had undertaken the defence of Philosophical Liberty, and the correspondence was closed without my being convinced of

the

the fallacy of my arguments, though upon studying the subject regularly, in the course of my academical education afterwards, I became a confirmed Necessarian, and I have through life derived, as I imagine, the greatest advantage from my full persuasion of the truth of that doctrine.

My Aunt, and all my relations, being strict Calvinists, it was their intention to send me to the academy at *Mile-end*, then under the care of Dr. Cawder. But, being at that time an Arminian, I resolutely opposed it, especially upon finding that if I went thither, besides giving an *experience*, I must subscribe my assent to ten printed articles of the strictest calvinistic faith, and repeat it every six months. My opposition, however, would probable have been to no purpose, and I must have adopted some other mode of life, if Mr. Kirkby above mentioned had not interposed, and strongly recommended the academy of Dr. Doddridge, on the idea that I should have a better chance of being made a scholar. He had received a good education himself, was a good classical scholar, and had no opinion of the mode of education among the very orthodox Dissenters, and being fond of me, he was

desirous

desirous of my having every advantage that could be procured for me. My good Aunt, not being a bigotted Calvinist, entered into his views, and Dr. Doddridge being dead, I was sent to Daventry, and was the first pupil that entered there. My Step-mother also, who was a woman of good sense, as well as of religion, had a high opinion of Dr. Doddridge, having been sometime housekeeper in his family. She had always recommended his Academy, but died before I went thither.

Three years, viz. from September 1752 to 1755, I spent at Daventry with that peculiar satisfaction with which young persons of generous minds usually go through a course of liberal study, in the society of others engaged in the same pursuits, and free from the cares and anxieties which seldom fail to lay hold on them when they come out into the world.

In my time, the academy was in a state peculiarly favorable to the serious pursuit of truth, as the students were about equally divided upon every question of much importance, such as Liberty and Necessity, the Sleep of the soul, and all the articles of theological orthodoxy and heresy; in consequence of which

all

all these topics were the subject of continual discussion. Our tutors also were of different opinions,
Dr. Ashworth taking the orthodox side of every
question, and Mr. Clark, the sub-tutor, that of heresy, though always with the greatest modesty.

Both of our tutors being young, at least as tutors,
and some of the senior students excelling more than
they could pretend to do in several branches of study, they indulged us in the greatest freedoms, so
that our lectures had often the air of friendly conversations on the subjects to which they related. We
were permitted to ask whatever questions, and to
make whatever remarks, we pleased; and we did
it with the greatest, but without any offensive, freedom. The general plan of our studies, which may
be seen in Dr. Doddridge's published lectures, was
exceedingly favourable to free enquiry; as we were
referred to authors on both sides of every question,
and were even required to give an account of them.
It was also expected that we should abridge the most
important of them for our future use. The public
library contained all the books to which we were
referred.

It was a reference to Dr. Hartley's Observations

on

on Man in the course of our Lectures, that first brought me acquainted with that performance, which immediately engaged my closest attention, and produced the greatest, and in my opinion the most favourable effect on my general turn of thinking thro' life. It established me in the belief of the doctrine of Necessity, which I first learned from Collins; it greatly improved that disposition to piety which I brought to the academy, and freed it from that rigour with which it had been tinctured. Indeed, I do not know whether the consideration of Dr. Hartley's theory contributes more to enlighten the mind, or improve the heart; it effects both in so super-eminent a degree.

In this situation, I saw reason to embrace what is generally called the heterodox side of almost every question.* But notwithstanding this, and though Dr.

It will be seen in the course of these memoirs that from time to time as deeper reflection and more extensive reading incited him, he saw reason to give up almost all the peculiar theological and metaphysical opinions which he had imbibed in early youth; some of them with considerable difficulty, and all of them at the evident risk of considerable obloquy from those whom he highly respected, as well as from those on whom his interest appeared to depend. T. C.

Dr. Ashworth was earnestly desirous to make me as orthodox as possible, yet, as my behaviour was unexceptionable, and as I generally took his part in some little things by which he often drew upon himself the ill-will of many of the students, I was upon the whole a favourite with him. I kept up more or less of a correspondence with Dr. Ashworth till the time of his death, though much more so with Mr. Clark. This continued till the very week of his melancholy death by a fall from his horse at Birmingham, where he was minister.

Notwithstanding the great freedom of our speculations and debates, the extreme of heresy among us was Arianism; and all of us, I believe, left the academy with a belief, more or less qualified, of the doctrine of *atonement*.

Warm friendships never fail to be contracted at places of liberal education; and when they are well chosen are of singular use; Such was mine with Mr. Alexander of Birmingham. We were in the same class, and during the first year occupied the same room. By engagements between ourselves we rose early, and dispatched many articles of business every day. One of them, which continued all the

time

time we were at the academy, was to read every day
ten folio pages in some Greek author, and generally
a Greek play in the course of the week besides. By
this means we became very well acquainted with
that language, and with the most valuable authors in
it. This exercise we continued long after we left
the academy, communicating to each other by letter
an account of what we read. My life becoming
more occupied than his, he continued his application
to Greek longer than I did, so that before his death
he was, I imagine, one of the best Greek scholars in
this or any other country. My attention was always
more drawn to mathematical and philosophical studies than his was.

These voluntary engagements were the more necessary, in the course of our academical studies, as
there was then no provision made for teaching the
learned languages. We had even no compositions,
or orations, in Latin. Our course of lectures was
also defective in containing no lectures on the scriptures, or on ecclesiastical history, aud by the students in general (and Mr. Alexander and myself
were no exceptions) commentators in general and
ecclesiastical history also, were held in contempt.

On leaving the academy he went to study under his uncle Dr. Benson, and with him learned to value the critical study of the scriptures so much, that at length he almost confined his attention to them.

My other particular friends among my fellow students were Mr. Henry Holland, of my own class, Messrs. Whitehead, Smithson, Rotherham, and Scholefield in that above me, and Mr. Taylor in that below me. With all these I kept up more or less of a correspondence, and our friendship was terminated only by the death of those who are now dead, viz. the three first named of these six, and I hope it will subsist to the same period with those who now survive.

All the while I was at the academy I never lost sight of the great object of my studies, which was the duties of a christian minister, and there it was that I laid the general plan which I have executed since. Particularly I there composed the first copy of my *Institutes of Natural and Revealed Religion*, Mr. Clark, to whom I communicated my scheme, carefully perusing every section of it, and talking over the subject of it with me.

But I was much discouraged even then with the
impedi-

impediment in my speech, which I inherited from my family, and which still attends me. Sometimes I absolutely stammered, and my anxiety about it was the cause of much distress to me. However, like St. Paul's *thorn in the flesh,* I hope it has not been without its use. Without some such check as this, I might have been disputatious in company, or might have been seduced by the love of popular applause as a preacher: whereas my conversation and my deliverery in the pulpit having nothing in them that was generally striking, I hope I have been more attentive to qualifications of a superior kind.

It is not, I believe, usual for young persons in dissenting academies to think much of their future situations in life. Indeed, we are happily precluded from that by the impossibility of succeeding in any application for particular places. We often, indeed, amused ourselves with the idea of our dispersion in all parts of the kingdom after living so happily together; and used to propose plans of meeting at certain times, and smile at the different appearance we should probably make after being ten or twenty years settled in the world. But nothing of this kind was ever seriously resolved upon by us.

For

For my own part, I can truly say I had very little ambition, except to distinguish myself by my application to the studies proper to my profession ; and I cheerfully listened to the first proposal that my tutor made to me, in consequence of an application made to him, to provide a minister for the people of Needham Market in Suffolk, though it was very remote from my friends in Yorkshire, and a very inconsiderable place.

When I went to preach at Needham as a candidate, I found a small congregation, about an hundred people, under a Mr. Meadows, who was superannuated   They had been without a minister the preceding year, on account of the smallness of the salary; but there being some respectable and agreeable families among them, I flattered myself that I should be useful and happy in the place, and therefore accepted the unanimous invitation to be assistant to Mr. Meadows, with a view to succeed him when he died.   He was a man of some fortune.

This congregation had been used to receive assistance from both the Presbyterian and Independent funds; but upon my telling them that I did not chuse to have any thing to do with the Independents,

and

and asking them whether they were able to make up
the salary they promised me (which was forty pounds
per annum) without any aid from the latter fund,
they assured me they could. I soon, however,
found that they deceived themselves ; for the most
that I ever received from them was in the proportion
of about thirty pounds per annum, when the expence of my board exceeded twenty pounds.

Notwithstanding this, every thing else for the first
half year appeared very promising, and I was happy
in the success of my schemes for promoting the interest of religion in the place. I catechised the children, though there were not many, using Dr. Watt's
Catechism ; and I opened my lectures on the theory
of religion from the *institutes*, which I had composed
at the academy, admitting all persons to attend them
without distinction of sex or age ; but in this I soon
found that I had acted imprudently. A minister in
that neighbourghood had been obliged to leave his
place on account of Arianism, and though nothing
had been said to me on the subject, and from the
people so readily consenting to give up the independent fund, I thought they could not have much bigotry among them, I found that when I came to
treat

treat of the *Unity of God*, merely as an article of religion, several of my audience were attentive to nothing but the soundness of my faith in the doctrine of the Trinity.

Also, though I had made it a rule to myself to introduce nothing that could lead to controversy into the pulpit; yet making no secret of my real opinions in conversation, it was soon found that I was an Arian. From the time of this discovery my hearers fell off apace, especially as the old minister took a decided part against me. The principal families, however, still continued with me; but notwithstanding this, my salary fell far short of thirty pounds per annum, and if it had not been for Dr. Benson and Dr. Kippis, especially the former, procuring me now and then an extraordinary five pounds from different charities, I do not believe that I could have subsisted. I shall always remember their kindness to me, at a time when I stood in so much need of it.

When I was in this situation, a neighbouring minister whose intimate friend had conformed to the church of England, talked to me on that subject. He himself, I perceived, had no great objection to it, but rejecting the proposal, as a thing that I could not think of, he never mentioned it to me any more.

To

To these difficulties, arising from the sentiments of my congregation, was added that of the failure of all remittances from my aunt, owing in part to the ill offices of my orthodox relations; but chiefly to her being exhausted by her liberality to others, and think_ing that when I was settled in the world, I ought to be no longer burdensome to her. Together with me she had brought up a niece, who was almost her on-ly companion, and being deformed, could not have subsisted without the greatest part, at least, of all she had to bequeath. In consequence of these circum-stances, tho' my aunt had always assured me that, if I chose to be a minister, she would leave me indepen dent of the profession, I was satisfied she was not able to perform her promise, and freely consented to her leaving all she had to my cousin ; I had only a silver tankard as a token of her remembrance. She had spared no expence in my education, and that was do-ing more for me than giving me an estate.

But what contributed greatly to my distress was the *impediment in my speech*, which had increased so much as to make preaching very painful, and took from me all chance of recommending myself to any better place. In this state, hearing of the proposal of

one

one Mr. Angier to cure all defects of speech, I prevailed upon my aunt to enable me to pay his price,
which was twenty guineas; and this was the first occasion of my visiting London.   Accordingly, I attended him about a month, taking an oath not to re-
veal his method, and I received some temporary benefit; but soon relapsed again, and spoke worse than
ever.   When I went to London it was in company
with Mr. Smithson, who was settled at Harlestown
in Norfolk.   By him I was introduced to Dr. Kippis
and Dr. Benson, and by the latter to Dr. Price, but
not at that time.

At Needham I felt the effect of a low despised situation, together with that arising from the want of popular talents.   There were several vacancies in congregations in that neighbourhood, where my sentiments would have been no objection to me, but I was
never thought of.   Even my next neighbours, whose
sentiments were as free as my own, and known to be
so, declined making exchanges with me, which,
when I left that part of the country, he acknowledged
was not owing to any dislike his people had to me as
heretical, but for other reasons, the more genteel part
of his hearers always absenting themselves when they
heard

heard I was to preach for him. But visiting that country some years afterwards, when I had raised myself to some degree of notice in the world, and being invited to preach in that very pulpit, the same people crowded to hear me, though my elocution was not much improved, and they professed to admire one of the same discourses they had formerly despised.

Notwithstanding these unfavorable circumstances, I was far from being unhappy at Needham. I was boarded in a family from which I received much satisfaction, I firmly believed that a wise providence was disposing every thing for the best, and I applied with great assiduity to my studies, which were classical, mathematical and theological. These required but few books. As to Experimental Philosophy, I had always cultivated an acquaintance with it, but I had not the means of prosecuting it.

With respect to miscellaneous reading, I was pretty well supplied by means of a library belonging to Mr. S. Alexander, a quaker,* to which I had the freest access.

---

* QUAKERS. That instances of liberality of sentiment with respect to religious opinion are frequently to be found among the Quakers there can be no doubt, but this is certainly no part of their character.

access. Here it was that I was first acquainted with any person of that persuasion; and I must acknowledge my obligation to many of them in every future stage of my life. I have met with the noblest instances of liberality of sentiment and the truest generosity among them.

My studies however, were chiefly theological. Having left the academy, as I have observed, with a qualified belief of the doctrine of *Atonement*, such as is found in Mr. Tomkin's book, entitled, *Jesus Christ*

the

---

as a Sect. Thomas Letchworth one of the most acute and ingenious of their preachers at Wandsworth near London, who from the writings of Dr. Priestley had become a firm convert to his Unitarian opinions, informed me that the expression of those opinions would be attended with certain expulsion from the Society. Very lately Hannah Bernard a female public friend who went from America to England, was prohibited from preaching by the Society, on account of her Unitarian doctrines.

Thomas Letchworth has been dead many years. In the short contest on the question of liberty and necessity which was occasioned by Toplady's life of Jerome Zanchius, he wrote a good defence of the doctrine of necessity signed Philaretes in answer to one from a disciple of Fletcher's of Madely, under the signature of Philaleutheros. There is a trifling account of him containing no information, by one William Matthews.　　　　　　　　　　　　　　T. C.

the Mediator, I was desirous of getting some more definite ideas on the subject, and with that view set myself to peruse the whole of the old and new testament, and to collect from them all the texts that appeared to me to have any relation to the subject. This I therefore did with the greatest care, arranging them under a great variety of heads. At the same time I did not fail to note such *general considerations* as occurred to me while I was thus employed. The consequence of this was, what I had no apprehension of when I began the work, viz. a full persuasion that the doctrine of Atonement, even in its most qualified sense, had no countenance either from scripture or reason. Satisfied of this, I proceeded to digest my observations into a regular treatise, which a friend of mine, without mentioning my name, submitted to the perusal of Dr. Fleming and Dr. Lardner. In consequence of this, I was urged by them to publish the greater part of what I had written. But being then about to leave Needham, I desired them to do whatever they thought proper with respect to it, and they published about half of my piece, under the title of the *Doctrine of Remission, &c.*

This circumstance introduced me to the acquaint-

ance

ance of Dr. Lardner, whom I always called upon when I visited London.   The last time I saw him, which was little more than a year before his death, having by letter requested him to give me some assistance with respect to the history I then prepared to write of the Corruptions of Christianity, and especially that article of it, he took down a large bundle of pamphlets, and turning them over at length shewing me my own; said, " This contains my sentiments on the subject."   He had then forgot that I wrote it, and on my remarking it, he shook his head, and said that his memory began to fail him; and that he had taken me for another person.   He was then at the advanced age of ninety one.   This anecdote is trifling in itself, but it relates to a great and good man,

I have observed that Dr. Lardner only wished to publish a part of the treatise which my friend put into his hand.   The other part of it contained remarks on the reasoning of the apostle of Paul, which he could not by any means approve.   They were, therefore, omitted in this publication.   But the attention which I gave to the writings of this apostle at the time that I examined them, in order to collect
                                                             passa-

passages relating to the doctrine of atonement, satisfied me that his reasoning was in many places far
from being conclusive; and in a separate work I examined every passage in which his reasoning appeared to me to be defective, or his conclusions ill supported; and I thought them to be pretty numerous.

At that time I had not read any commentary on the
scriptures, except that of Mr. Henry when I was
young. However, seeing so much reason to be dissatisfied with the apostle Paul as a reasoner, I read
*Dr. Taylor's paraphrase on the epistle to the Ro-
mans;* but it gave me no sort of satisfaction; and
his general *Key to the epistles* still less. I therefore
at that time wrote some remarks on it, which were a
long time after published in the *Theological Repository* Vol. 4.

As I found that Dr. Lardner did not at all relish
any of my observations on the imperfections of the
sacred writers, I did not put this treatise into his
hands; but I shewed it to some of my younger
friends, and also to Dr. Kippis; and he advised me
to publish it under the character of an unbeliever,
in order to draw the more attention to it. This I

did

did not chuse, having always had a great aversion to assume any character that was not my own, even so much as disputing for the sake of discovering truth. I cannot ever say that I was quite reconciled to the idea of writing to a fictitious person, as in my *letters to a philosophical unbeliever*, though nothing can be more innocent, or sometimes more proper; our Saviour's parables implying a much greater departure from strict truth than those letters do. I therefore wrote the book with great freedom, indeed, but as a christian, and an admirer of the apostle Paul, as I always was in other respects.

When I was at Nantwich I sent this treatise to the press; but when nine sheets were printed off, Dr. Kippis dissuaded me from proceeding, or from publishing any thing of the kind, until I should be more known, and my character better established. I therefore desisted.; but when I opened the theological Repository, I inserted in that work every thing that was of much consequence in the other, in order to its being submitted to the examination of learned christians. Accordingly these communications were particularly animadverted upon by Mr. Willet of Neweastle, under the signature of W. W. But

I can-

I cannot say that his remarks gave me much satis-faction.

When I was at Needham I likewise drew up a treatise on the doctrine of *divine influence*, having collected a number of texts for that purpose, and ar-ranged them under proper heads, as I had done those relating to the doctrine of atonement. But I pub-lished nothing relating to it until I made use of some of the observations in my *sermon* on that subject, delivered at an ordination, and published many years afterwards.

While I was in this retired situation, I had, in con-sequence of much pains and thought, become per-suaded of the falsity of the doctrine of atonement, of the inspiration of the authors of the books of scripture as writers, and of all idea of supernatural influence, except for the purpose of miracles. But I was still an Arian, having never turned my attention to the Socinian doctrine, aud contenting myself with seeing the absurdity of the trinitarian system.

Another task that I imposed on myself, and in part executed at Needham, was an accurate compa-rison of the Hebrew text of the hagiographa and the prophets with the version of the Septuagint, noting

all

all the variations, &c.   This I had about half finished before I left that place ; and I never resumed it, except to do that occasionally for particular passages, which I then began, though with many disadvantages, with a design to go through the whole.   I had no Polyglot Bible, and could have little help from the labours of others.

The most learned of my acquaintance in this situation was Mr. Scott of Ipswich, who was well versed in the Oriental languages, especially the Arabic. But though he was far from being Calvinistical, he gave me no encouragement in the very free enquiries which I then entered upon.   Being excluded from all communication with the more orthodox ministers in that part of the country, all my acquaintance among the dissenting ministers, besides Mr. Scott, were Mr. Taylor of Stow-market, Mr. Dickinson of Diss, and Mr. Smithson of Harlestone ; and it is rather remarkable, that we all left that country in the course of the same year ; Mr. Taylor removing to Carter's lane in London, Mr. Dickinson to Sheffield, and Mr. Smithson to Nottingham.

But I was very happy in a great degree of intimacy with Mr. Chauvet, the rector of Stow-market.

He

He was descended of French parents; and I think was not born in England. Whilst he lived we were never long without seeing each other. But he was subject to great unevenness of spirits, sometimes the most chearful man living, and at other times most deplorably low. In one of these fits he at length put an end to his life. I heard afterwards that he had at one time been confined for insanity, and had even made the same attempt some time before.

Like most other young men of a liberal education, I had conceived a great aversion to the business of a schoolmaster, and had often said, that I would have recourse to any thing else for a maintenance in preference to it. But having no other resource, I was at length compelled by necessity to make some attempt in that way; and for this purpose I printed and distributed *Proposals,* but without any effect. Not that I was thought to be unqualified for this employment, but because I was not orthodox. I had proposed to teach the classics, mathematics, &c. for half a guinea per quarter, and to board the pupils in the house with myself for twelve guineas per annum.

Finding this scheme not to answer, I proposed to

give lectures to grown persons in such branches of sciencs as I could conveniently procure the means of doing; and I began with reading about twelve lectures on the *use of the Globes*, at half a guinea.   I had one course of ten hearers, which did something more than pay for my globes; and I should have proceeded in this way, adding to my apparatus as I should have been able to afford it, if I had not left that place, which was in the following manner.

My situation being well known to my friends, Mr. Gill, a distant relation by my mother, who had taken much notice of me before I went to the academy, and had often lent me books, procured me an invitation to preach as a candidate at Sheffield, on the resignation of Mr. Wadsworth.   Accordingly I did preach as a candidate   but though my opinions were no objection to me there, I was not approved.   But Mr. Haynes, the other minister, perceiving that I had no chance at Sheffield, told me that he could recommend me to a congregation at Nantwich in Cheshire, where he himself had been settled ; and as it was at a great distance from Needham he would endeavour to procure me an invitation to preach there for a year certain.   This he did, and I gladly accepting of it, removed

ved from Needham , going thence to London by sea,
to save expence.   This was in 1758, after having
been at Needham just three years.*

At

---

* It is about sixty miles from Needham to London, so that the roads
must have been in a bad state to render a water passage more eligible
than by land.   The first turnpike in England was authorized by an
act of Ch. II. 1663 but the system was not adopted with spirit until
near the middle of the last century.   The manufacturing inland towns
of Great Britain, such as Manchester, Leeds, Halifax, &c. chiefly
carried on their business through the medium of travelling pedlars,
and afterwards on pack horses.   The journey in this manner from
Manchester to London occupied a fortnight ; and it was not unusual
for a trader going the first time himself on this expedition to take the
prudent precaution of making his will.   At present the mail stage per-
forms the journey in about a day and a half.   In the beginning of this
century (as Dr. Aikin in his history of Manchester observes) it was
thought a most arduous undertaking to make a public road over the
hills that separate Yorkshire and Lancashire ; now, they are pierced
by three navigable canals.   Indeed the prosperous state of British
manufactures and commerce, seems to have originated and progressed
with the adoption of turnpikes and canals.   They facilitate not merely
the carriage and interchange of heavy materials necessary to machine-
ry, but they make personal intercourse cheap, speedy and universal ;
they thus furnish the means of seeing and communicating improvements, and of observing in what way one manufacture may be brought
to bear upon another widely different in its kind,   We are not yet
sufficiently aware of their importance in America, even to the interests
of agriculture.   T. C.

At Nantwich I found a good natured friendly people, with whom I lived three years very happily; and in this situation I heard nothing of those controversies which had been the topics of almost every conversation in Suffolk; and the consequence was that I gave little attention to them myself. Indeed it was hardly in my power to do it, on account of my engagement with a school, which I was soon able to establish, and to which I gave almost all my attention; and in this employment, contrary to my expectations, I found the greatest satisfaction, notwithstanding the confinement and labour attending it.

My school generally consisted of about thirty boys, and I had a separate room for about half a dozen young ladies. Thus I was employed from seven in the morning untill four in the afternoon, without any interval except one hour for dinner, and I never gave a holiday on any consideration, the red letter days, as they are called, excepted. Immediately after this employment in my own school rooms, I went to teach in the family of Mr. Tomkinson, an eminent attorney, and a man of large fortune, whose recommendation was of the greatest service to me; and here I continued until seven in the evening. I had therefore but

little

little leisure for reading or for improving myself in
any way, except what necessarily arose from my employment.

Being engaged in the business of a schoolmaster,
I made it my study to regulate it in the best manner,
and I think I may say with truth, that in no school
was more business done, or with more satisfaction,
either to the master, or the scholars, than in this of
mine. Many of my scholars are probably living and
I am confident that they will say that this is no vain
boast.

At Needham I was barely able with the greatest e-
conomy to keep out of debt (though this I always made
a point of doing at all events) but at Nantwich my
school soon enabled me to purchase a few books, and
some philosophical instruments, as a small air pump,
an electrical machine, &c. These I taught my scholars in the highest class to keep in order, and make
use of, and by entertaining their parents and friends
with experiments, in which the scholars were generally the operators, and sometimes the lecturers too,
I considerably extended the reputation of my school;
though I had no other object originally than gratifying
my own taste. I had no leisure, however, to make

any

any original experiments until many years after this time.

As there were few children in the congregation (which did not consist of more than sixty persons, and a great proportion of them travelling scotchmen) there was no scope for exertion with respect to my duty as a minister. I therefore contented myself with giving the people what assistance I could at their own houses, where there were young persons; and I added very few sermons to those which I had composed at Needham, where I never failed to make at least one every week.

Being boarded with Mr. Eddowes, a very sociable and sensible man, and at the same time the person of the greatest property in the congregation, and who was fond of music, I was induced to learn to play a little on the English flute, as the easiest instrument; and though I was never a proficient in it, my playing contributed more or less to my amusement many years of my life. I would recommend the knowledge and practice of music to all studious persons; and it will be better for them, if, like myself, they should have no very fine ear, or exquisite taste; as by this means they will be more easily

pleased,

pleased, and be less apt to be offended when the performances they hear are but indifferent.

At Nantwich I had hardly any literary acquaintance besides Mr. Brereton, a clergyman in the neighbourhood, who had a taste for astronomy, philosophy, and literature in general. I often slept at his house, in a room to which he gave my name. But his conduct afterwards was unworthy of his profession.

Of dissenting ministers I saw most of Mr. Keay of Whitchurch, and Dr. Harwood, who lived and had a school at Congleton, preaching alternately at Leek and Wheelock, the latter place about ten miles from Nantwich. Being both of us schoolmasters, and having in some respect the same pursuits, we made exchanges for the sake of spending a Sunday evening together every six weeks in the summer time. He was a good classical scholar, and a very entertaining companion.

In my congregation there was (out of the house in which I was boarded) hardly more than one family in which I could spend a leisure hour with much satisfaction, and that was Mr. James Caldwall's, a scotchman. Indeed, several of the travelling

ling Scotchmen who frequented the place, but made no long stay at any time, were men of very good sense; and what I thought extraordinary, not one of them was at all Calvinistical.

My engagements in teaching allowed me but little time for composing any thing while I was at Nantwich. There, however, I recomposed my *Observations on the character and reasoning of the apostle Paul*, as mentioned before. For the use of my school I then wrote an *English grammer** on a new plan, leaving out all such technical terms as were borrowed from other languages, and had no corresponding modifications in ours, as the future tense, &c. and to this I afterwards subjoined *Observations for the use of proficients in the language*,‖ from the notes which I collected at Warrington; where, being tutor in the languages and Belles Letters, I gave particular attention to the English language, and intended

---

* Printed in 1761.

‖ Printed in 1772 at London. His lectures on the Theory of Language and Universal Grammar were printed the same year at Warrington. David Hume was made sensible of the Gallicisms and Peculiarities of his stile by reading this Grammar; He acknowledged it to Mr. Griffith the Bookseller, who mentioned it to my father.

tended to have composed a large treatise on the structure and present state of it. But dropping the scheme in another situation, I lately gave such parts of my collection as I had made no use of to Mr. Herbert Croft of Oxford, on his communicating to me his design of compiling a Dictionary and Grammar of our language.

The academy at Warrington was instituted when I was at Needham, and Mr. Clark knowing the attention that I had given to the learned languages when I was at Daventry, had then joined with Dr. Benson and Dr. Taylor in recommending me as tutor in the languages. But Mr. (afterward Dr.) Aikin, whose qualifications were superior to mine, was justly prefered to me. However, on the death of Dr. Taylor, and the advancement of Mr. Aikin to be tutor in divinity, I was invited to succeed him. This I accepted, though my school promised to be more gainful to me. But my employment at Warrington would be more liberal, and less painful. It was also a means of extending my connections. But, as I told the persons who brought me the invitation, viz. Mr. Seddon and Mr. Holland of Bolton, I should have preferred the office of teaching the mathematics and

natural

natural philosophy, for which I had at that time a great predilection.

My removal to Warrington was in September, 1761, after a residence of just three years at Nantwich. In this new situation I continued six years, and in the second year I married a daughter of Mr. Isaac Wilkinson, an Ironmaster near Wrexham in Wales, with whose family I had became acquainted in consequence of having the youngest son, William, at my school at Nantwich. This proved a very suitable and happy connection, my wife being a woman of an excellent understanding, much improved by reading, of great fortitude and strength of mind, and of a temper in the highest degree affectionate and generous ; feeling strongly for others, and little for herself. Also, greatly excelling in every thing relating to household affairs, she entirely relieved me of all concern of that kind, which allowed me to give all my time to the prosecution of my studies, and the other duties of my station. And though, in consequence of her father becoming impoverished, and wholly dependent on his children, in the latter part of his life, I had little fortune with her, I unexpectedly found a great resource in her two brothers, who had become wealthy,

wealthy, especially the elder of them. At Warrington I had a daughter, Sarah, who was afterwards married to Mr. William Finch of Heath forge near Dudley.

Though at the time of my removal to Warrington I had no particular fondness for the studies relating to my profession then, I applied to them with great assiduity; and besides composing courses of *Lectures on the theory of Language*, and on *Oratory and Criticism*, on which my predecessor had lectured, I introduced lectures on *history and general policy*, on the *laws and constitutions of England*, and on the *history of England*. This I did in consequence of observing that, though most of our pupils were young men designed for situations in civil and active life, every article in the plan of their education was adapted to the learned professions.

In order to recommend such studies as I introduced, I composed an *essay on a course of liberal education for civil and active life*, with *syllabuses* of my three new courses of lectures; and Dr. Brown having just then published a plan of education, in which he recommended it to be undertaken by the state, I added some *remarks on his treatise*, shewing how inimical

mical it was to liberty, and the natural rights of parents.   This leading me to consider the subject of
civil and political liberty, I published my thoughts
on it, in an *essay on government*, which in a second
edition I much enlarged, including in it what I wrote
in answer to Dr. Balguy, on church authority, as
well as my animadversions on Dr. Brown.

My *Lectures on the theory of language and universal
grammar* were printed for the use of the students, but
they were not published.   Those on *Oratory and
Criticism* I published when I was with Lord Shel
burne, and those on *History and general policy* are
now printed, and about to be published.*

Finding no public exercises at Warrington, I introduced them there, so that afterwards every Saturday the tutors, all the students, and often strangers,
were assembled to hear English and Latin compositions, and sometimes to hear the delivery of speeches,
and the exhibition of scenes in plays.   It was my
province to teach elocution, and also Logic, and
Hebrew.   The first of these I retained; but after a
year

---

* This work has been reprinted in Philadelphia with additions, particularly of a chapter on the government of the United States.

year or two I exchanged the two last articles with
Dr. Aikin for the civil law, and one year I gave a
course of lectures in anatomy.

With a view to lead the students to a facility in
writing English, I encouraged them to write in verse.
This I did not with any design to make them poets,
but to give them a greater facility in writing prose ,
and this method I would recommend to all tutors.
I was myself far from having any pretension to the
character of a poet ; but in the early part of my life I
was a great versifier, and this, I believe, as well as
my custom of writing after preachers, mentioned before, contributed to the ease with which I always
wrote prose.   Mrs. Barbauld has told me that it was
the perusal of some verses of mine that first induced
her to write any thing in verse, so that this country is
in some measure indebted to me for one of the best
poets it can boast of.   Several of her first poems
were written when she was in my house, on occasions that occurred while she was there.

It was while I was at Warrington that I published
my *Chart of Biography*, though I had begun to con-
struct it at Nantwich.   Lord Willoughby of Parham,
who lived in Lancashire, being pleased with the idea

D   of

of it, I, with his consent, inscribed it to him; but he died before the publication of it : The *Chart of History*, corresponding to it, I drew up some time after at Leeds.

I was in this situation when, going to London,* and being introduced to Dr. Price, Mr. Canton, Dr. Watson, (the Physician,) and Dr. Franklin, I was led to attend to the subject of experimental philosophy more than I had done before ; and having composed all the Lectures I had occasion to deliver and finding myself at liberty for any undertaking, I mentioned to Dr. Franklin an idea that had occurred to me of writing the history of discoveries in Electricity, which had been his favourite study. This I told him might be an useful work, and that I would willingly undertake it, provided I could be furnished with the books necessary for the purpose. This he readily undertook, and my other friends assisting him in it, I set about the work, without having the

least

---

* He always spent one month in every year in London which was of great use to him. He saw and heard a great deal. He generally made additions to his library and his chemical apparatus. A new turn was frequently given to his ideas. New and useful acquaintances were formed, and old ones confirmed.

least idea of doing any thing more than writing a distinct and methodical account of all that had been done by others. Having, however, a pretty good machine, I was led, in the course of my writing the history, to endeavour to ascertain several facts which were disputed; and this led me by degrees into a large field of original experiments, in which I spared no expence that I could possibly furnish.

These experiments employed a great proportion of my leisure time; and yet before the complete expiration of the year in which I gave the plan of my work to Dr. Franklin, I sent him a copy of it in print. In the same year five hours of every day were employed in lectures, public or private, and one two months vacation I spent chiefly at Bristol, on a visit to my father-in-law.

This I do not mention as a subject of boasting. For many persons have done more in the same time; but as an answer to those who have objected to some of my later writings, as hasty performances. For none of my publications were better received than this *History of Electricity*, which was the most hasty of them all. However, whether my publications have taken up more or less time, I am confident that

more

more would not have contributed to their perfection, in any essential particular; and about anything farther I have never been very solicitous. My object was not to acquire the character of a fine writer, but of an useful one. I can also truly say that gain was never the chief object of any of my publications. Several of them were written with the prospect of certain loss.

During the course of my electrical experiments in this year I kept up a constant correspondence with Dr. Franklin, and the rest of my philosophical friends in London; and my letters circulated among them all, as also every part of my History as it was transcribed. This correspondence would have made a considerable volume, and it took up much time; but it was of great use with respect to the accuracy of my experiments, and the perfection of my work.

After the publication of my Chart of Biography, Dr. Percival of Manchester, then a student at Edinburgh, procured me the title of Doctor of laws from that university; and not long after my new experiments in electricity were the means of introducing me into the Royal Society, with the recommendation of Dr. Franklin, Dr. Watson, Mr. Canton, and Dr. Price.

In

In the whole time of my being at Warrington I was singularly happy in the society of my fellow tutors,* and of Mr. Seddon, the minister of the place. We drank tea together every Saturday, and our conversation was equally instructive and pleasing. I often thought it not a little extraordinary, that four persons, who had no previous knowledge of each other, should have been brought to unite in conducting such a scheme as this, and all be zealous necessarians, as we were. We were likewise all Arians, and the only subject of much consequence on which we differed respected the doctrine of atonement, concerning which Dr. Aikin held some obscure notions. Accordingly, this was frequently the topic of our friendly conversations. The only Socinian in the neighbourhood was Mr. Seddon of Manchester; and we all wondered at him. But then we never entered into any particular examination of the subject.

Receiving some of the pupils into my own house,

I was

---

* At Warrington he had for colleagues and successors, Dr. John Taylor, author of the Hebrew Concordance and of several other works, on Original Sin, Atonement, &c. Dr. Aikin the Elder, Dr. Reinhold Forster the Naturalist and traveller, Dr. Enfield and Mr. Walker.

I was by this means led to form some valuable friendships, but especially with Mr. Samuel Vaughan, a friendship which has continued hitherto, has in a manner connected our families, and will, I doubt not, continue through life. The two eldest of his sons were boarded with me.

The tutors having sufficient society among themselves, we had not much acquaintance out of the academy. Sometimes, however, I made an excursion to the towns in the neighbourhood. At Liverpool I was always received by Mr. Bentley, afterwards partner with Mr. Wedgwood, a man of excellent taste improved understanding, and a good disposition, but an unbeliever in christianity, which was therefore often the subject of our conversation. He was then a widower, and we generally, and contrary to my usual custom, sat up late. At Manchester I was always the guest of Mr. Potter, whose son Thomas was boarded with me. He was one of the worthiest men that ever lived. At Chowbent I was much acquainted with Mr. Mort, a man equally distinguished by his chearfulness and liberality of sentiment.

Of the ministers in the neighbourhood, I recollect
with

with much satisfaction the interviews I had with Mr. Godwin of Gataker, Mr. Holland of Bolton, and Dr. Enfield of Liverpool, afterwards tutor at Warrington.

Though all the tutors in my time lived in the most perfect harmony, though we all exerted ourselves to the utmost, and there was no complaint of want of discipline, the academy did not flourish. There had been an unhappy difference between Dr. Taylor and the trustees, in consequence of which all his friends, who were numerous, were our enemies; and too many of the subscribers, being probably weary of the subscription, were willing to lay hold of any pretence for dropping it, and of justifying their conduct afterwards.

It is possible that in time we might have overcome the prejudices we laboured under, but there being no prospect of things being any better, and my wife having very bad health, on her account chiefly I wished for a removal, though nothing could be more agreeable to me at the time than the whole of my employment, and all the laborious part of it was over. The terms also on which we took boarders, viz. 15 £. per annum, and my salary being only

100 £. per annum with a house, it was not possible, even living with the greatest frugality, to make any provision for a family. I was there six years, most laboriously employed, for nothing more than a bare subsistence. I therefore listened to an invitation to take the charge of the congregation of Mill-hill chapel at Leeds, where I was pretty well known, and thither I removed in September 1767.

Though while I was at Warrington it was no part of my duty to preach, I had from choice continued the practice; and wishing to keep up the character of a dissenting minister, I chose to be ordained while I was there; and though I was far from having conquered my tendency to stammer, and probably never shall be able to do it effectually, I had, by taking much pains, improved my pronunciation some time before I left Nantwich; where for the two first years this impediment had increased so much, that I once informed the people, that I must give up the business of preaching, and confine myself to my school. However, by making a practice of reading very loud and very slow every day, I at length succeeded in getting in some measure the better of this defect, but I am still obliged occasionally to have recourse to the same expedient.

At

At Leeds I continued six years very happy with a liberal, friendly, and harmonious congregation, to whom my services (of which I was not sparing) were very acceptable. Here I had no unreasonable prejudices to contend with, so that I had full scope for every kind of exertion; and I can truly say that I always considered the office of a christian minister as the most honourable of any upon earth, and in the studies proper to it I always took the greatest pleasure.

In this situation I naturally resumed my application to speculative theology, which had occupied me at Needham, and which had been interrupted by the business of teaching at Nantwich and Warrington. By reading with care Dr. *Lardner's letter on the logos*, I became what is called a Socinian soon after my settlement at Leeds; and after giving the closest attention to the subject, I have seen more and more reason to be satisfied with that opinion to this day, and likewise to be more impressed with the idea of its importance.

On reading Mr. *Mann's Dissertation on the times of the birth and death of Christ*, I was convinced that he was right in his opinion of our Saviour's ministry

ministry having continued little more than one year, and on this plan I drew out a *Harmony of the gospels*, the outline of which I first published in the Theological Repository, and afterwards separately and at large, both in Greek and English, with Notes, and an occasional Paraphrase. In the same work I published my *Essay on the doctrine of Atonement*, improved from the tract published by Dr. Lardner, and also my animadversions on the reasoning of the apostle Paul.

The plan of this *Repository* occured to me on seeing some notes that Mr. Turner of Wakefield had drawn up on several passages of scripture, which I was concerned to think should be lost. He very much approved of my proposal of an occasional publication, for the purpose of preserving such original observations as could otherwise probably never see the light. Of this work I published three volumes while I was at Leeds, and he never failed to give me an article for every number of which they were composed.

Giving particular attention to the duties of my office, I wrote several tracts for the use of my congregation, as two *Catechisms*, an *Address to masters*

ters

ters of families on the subject of family prayer, a *discourse on the Lord's Supper*, and on *Church discipline*, and *Institutes of Natural and Revealed religion.* Here I formed three classes of Catechumens; and took great pleasure in instructing them in the principles of religion. In this respect I hope my example has been of use in other congregations.

The first of my controversial treatises was written here in reply to some angry remarks on my discourse on the Lord's Supper by Mr. Venn, a clergyman in the neighbourhood. I also wrote remarks on Dr. *Balguy's sermon on Church authority*, and on some paragraphs in Judge *Blackstone's Commentaries* relating to the dissenters. To the two former no reply was made; but to the last the judge replied in a small pamphlet; on which I addressed a letter to him in the St. James's Chronicle. This controversy led me to print another pamphlet, entitled *The Principles and Conduct of the Dissenters with respect to the civil and ecclesiastical constitution of this country.* With the encouragement of Dr. Price and Dr. Kippis, I also wrote an *Address to Protestant Dissenters as such;* but without my name. Several of these pamphlets having been animadverted

upon

upon by an anonymous acquaintance, who thought
I had laid too much stress on the principles of the
Dissenters, I wrote a defence of my conduct in *Let-*
ters addressed to him.

The methodists being very numerous in Leeds,
and many of the lower sort of my own hearers listen-
ing to them, I wrote an *Appeal to the serious profes-*
sors of Christianity, an *Illustration of particular texts,*
and republished the *Trial of Elwall,* all in the cheapest manner possible. Those small tracts had a great
effect in establishing my hearers in liberal principles
of religion, and in a short time had a far more extensive influence than I could have imagined. By this
time more than thirty thousand copies of the Appeal
have been dispersed.

Besides these theoretical and controversial pieces,
I wrote while I was at Leeds my *Essay on Govern-*
ment mentioned before, my *English Grammar* enlarged, a *familiar introduction to the study of electricity,*
a *treatise on perspective,* and my *Chart of History,* and
also some anonymous pieces in favour of civil liberty during the persecution of Mr. Wilkes, the principal of which was *An Address to Dissenters on the sub-*
ject of the difference with America, which I wrote at
                                                    the

the request of Dr. Franklin, and Dr. Fothergil.

But nothing of a nature foreign to the duties of my profession engaged my attention while I was at Leeds so much as the prosecution of my experiments relating to *electricity*, and especially the doctrine of *air*. The last I was led into in consequence of inhabiting a house adjoining to a public brewery, where I at first amused myself with making experiments on the fixed air which I found ready made in the process of fermentation. When I removed from that house, I was under the necessity of making the fixed air for myself; and one experiment leading to another, as I have distinctly and faithfully noted in my various publications on the subject, I by degrees contrived a convenient apparatus for the purpose, but of the cheapest kind.

When I began these experiments I knew very little of *chemistry*, and had in a manner no idea on the subject before I attended a course of chemical lectures delivered in the academy at Warrington by Dr. Turner* of Liverpool. But I have often thought

that

---

* Dr. TURNER was a Physician at Liverpool: among his friends a professed Atheist. It was Dr. Turner who wrote the reply to Dr.

Priestley's

that upon the whole, this circumstance was no disadvantage to me; as in this situation I was led to
devise an apparatus, and processes of my own, adapted to my peculiar views. Whereas, if I had been
previously accustomed to the usual chemical processes, I should not have so easily thought of any
other ; and without new modes of operation I should
hardly have discovered any thing materially new.*

                                                    My

Priestley's letters to a philosophical unbeliever under the feigned name
of Hammon. He was in his day a good practical chemist.   I believe
it was Dr. Turner who first invented, or at least brought to tolerable
perfection, the art of copying prints upon glass, by striking off impressions with a coloured solution of silver and fixing them on the glass by
baking on an iron plate in a heat sufficient to incorporate the solution
with the glass.   Some of them are very neatly performed, producing
transparent copies in a bright yellow upon the clear glass.

    Dr. Turner was not merely a whig but a republican.   In a friendly
debating society at Liverpool about the close of the American war, he
observed in reply to a speaker who had been descanting on the honour
Great Britain had gained during the reign of his present Majesty,
that it was true, we had lost the *Terra firma* of the thirteen colonies in
America, but we ought to be satisfied with having gained in return, by
the generalship of Dr. Herschel, a terra incognita of much greater
extent *in nubibus*.   T. C.

    * This necessary attention to economy also aided the simplicity of
his apparatus, and was the means in some degree of improving it in
                                                    this

My first publication on the subject of air was in 1772. It was a small pamphlet, on the method of impregnating water with fixed air ; which being immediately translated into French, excited a great degree of attention to the subject, and this was much increased by the publication of my first paper of experiments in a large article of the Philosophical Transactions the year following, for which I received the gold medal of the society. My method of impregnating water with fixed air was considered at a meeting of the College of Physicians, before whom I made the experiments, and by them it was recommended to the Lords of the Admiral (by whom they had been summoned for the purpose) as likely to be of use in the sea scurvy.

The only person in Leeds who gave much attention to my experiments was Mr. Hay, a surgeon. He was a zealous methodist, and wrote answers to some

of

---

this important respect. This plainness of his apparatus rendered his experiments easy to be repeated, and gave them accuracy. In this respect he was like his great Cotemporary Scheele, whose discoveries were made by means easy to be procured and at small expence. The French Chemists have adopted a practice quite the reverse. T. C.

of my theological tracts; but we always conversed with the greatest freedom on philosophical subjects, without mentioning any thing relating to theology. When I left Leeds, he begged of me the earthen trough in which I had made all my experiments on air while I was there. It was such an one as is there commonly used for washing linnen.

Having succeeded so well in the History of Electricity, I was induced to undertake the history of all the branches of experimental philosophy; and at Leeds I gave out proposals for that purpose, and published the *History of discoveries relating to vision light and colours.* This work, also, I believe I executed to general satisfaction, and being an undertaking of great expence, I was under the necessity of publishing it by subscription. The sale, however, was not such as to encourage me to proceed with a work of so much labour and expence; so that after purchasing a great number of books, to enable me to finish my undertaking, I was obliged to abandon it, and to apply wholly to original experiments.*

In writing the History of discoveries relating to vision,

---

* Many of the subscriptions remained unpaid.

vision, I was much assisted by Mr. Michell, the discoverer of the method of making artificial magnets. Living at Thornhill, not very far from Leeds, I frequently visited him, and was very happy in his society, as I also was in that of Mr. Smeaton, who lived still nearer to me. He made me a present of his excellent air pump, which I constantly use to this day. Having strongly recommended his construction of this instrument, it is now generally used; whereas before that hardly any had been made during the twenty years which had elapsed after the account that he had given of it in the Philosophical Transactions.

I was also instrumental in reviving the use of large electrical machines, and batteries, in electricity, the generality of electrical machines being little more than play things at the time that I began my experiments. The first very large electrical machine was made by Mr. Nairne in consequence of a request made to me by the Grand Duke of Tuscany, to get him the best machine that we could make in England. This, and another that he made for Mr. Vaughan, were constituted on a plan of my own. But afterwards Mr. Nairne made large machines on a more simple and improved construction ; and in consideration of

E

the service which I had rendered him, he made me a present of a pretty large machine of the same kind.

The review of my history of electricity by Mr. Bewley, who was acquainted with Mr. Michell, was the means of opening a correspondence between us, which was the source of much satisfaction to me as long as he lived. I instantly communicated to him an account of every new experiment that I made, and, in return, was favoured with his remarks upon them. All that he published of his own were articles in the *Appendixes* to my volumes on air, all of which are ingenious and valuable. Always publishing in this manner, he used to call himself my *satellite*. There was a vein of pleasant wit and humour in all his correspondence, which added greatly to the value of it. His letters to me would have made several volumes, and mine to him still more. When he found himself dangerously ill, he made a point of paying me a visit before he died; and he made a journey from Norfolk to Birmingham, accompanied by Mrs. Bewley, for that purpose; and after spending about a week with me, he went to his friend Dr. Burney, and at his house he died.

While I was at Leeds a proposal was made to me

to

to accompany Captain Cook in his second voyage to the south seas. As the terms were very advantageous, I consented to it, and the heads of my congregation had agreed to keep an assistant to supply my place during my absence. But Mr. Banks informed me that I was objected to by some clergymen in the board of longitude, who had the direction of this business, on account of my religious principles; and presently after I heard that Dr. Forster, a person far better qualified for the purpose, had got the appointment. As I had barely acquiesced in the proposal, this was no disappointment to me, and I was much better employed at home, even with respect to my philosophical pursuits. My knowledge of natural history was not sufficient for the undertaking; but at that time I should by application have been able to supply my deficiency, though now I am sensible I could not do it.

At Leeds I was particularly happy in my intercourse with Mr. Turner of Wakefield, and occasionally, with Mr. Cappe of York, and Mr. Graham of Halifax. And here it was that, in consequence of a visit which in company with Mr. Turner I made to the Archdeacon Blackburne at Richmond (with

E 2

whom

whom I had kept up a correspondence from the time that his son was under my care at Warrington) I first met with Mr. Lindsey, then of Catterick, and a correspondence and intimacy commenced, which has been the source of more real satisfaction to me than any other circumstance in my whole life. He soon discovered to me that he was uneasy in his situation, and had thoughts of quitting it. At first I was not forward to encourage him in it, but rather advised him to make what alteration he thought proper in the offices of the church, and leave it to his superiors to dismiss him if they chose. But his better judgment, and greater fortitude, led him to give up all connexion with the established church of his own accord.

This took place about the time of my leaving Leeds, and it was not until long after this that I was apprized of all the difficulties he had to struggle with before he could accomplish his purpose. But the opposition made to it by his nearest friends, and those who might have been expected to approve of the step that he took, and to have endeavoured to make it easy to him, was one of the greatest. Notwithstanding this he left Catterick, where he had lived

in

in affluence idolized by his parish, and went to London without any certain prospect; where he lived
in two rooms of a ground floor, until by the assistance of his friends, he was able to pay for the use of
the upper apartments, which the state of his health
rendered necessary.   In this humble situation have I
passed some of the most pleasing hours of my life,
when, in consequence of living with Lord Shelburne, I spent my winters in London.

On this occasion it was that my intimacy with
Mr. Lindsey was much improved, and an entire
concurrence in every thing that we thought to be for
the interest of christianity gave fresh warmth to our
friendship.   To his society I owe much of my zeal
for the doctrine of the divine unity, for which he
made so great sacrifices, and in the defence of which
he so much distinguished himself, so as to occasion
a new æra in the history of religion in this country.

As we became more intimate, confiding in his
better taste and judgment, and also in that of Mrs.
Lindsey, a woman of the same spirit and views, and
in all respects a help meet for him, I never chose to
publish any thing of moment relating to Theology
without consulting him; and hardly ever ventured

to insert any thing that they disapproved, being sensible that my disposition led to precipitancy, to which their coolness was a seasonable check.

At Leeds began my intercourse with Mr. Lee of Lincoln's Inn. He was a native of the place, and exactly one week older than myself. At that time he was particularly connected with the congregation, and before he was married spent his vacations with us. His friendship was a source of much greater satisfaction and advantage to me after I came to reside in London, and especially at the time of my leaving Lord Shelburne, when my prospects wore rather a cloudy aspect.

When I visited London, during my residence at Leeds, commenced my particular friendship for Dr. Price, to whom I had been introduced several years before by Dr. Benson; our first interview having been at Mr. Brownsword's at Newington, where they were members of a small literary society, in which they read various compositions. At that time Dr. Benson read a paper which afterwards made a section in his *Life of Christ*. For the most amiable simplicity of character, equalled only by that of Mr. Lindsey, a truly christian spirit, disinterested patriotism,

triotism, and true candour, no man in my opinion ever exceeded Dr. Price. His candour will appear the more extraordinary, considering his warm attachments to the theological sentiments which he embraced in very early life. I shall ever reflect upon our friendship as a circumstance highly honourable, as it was a source of peculiar satisfaction, to me.

I had two sons born to me at Leeds, Joseph and William, and though I was very happy there, I was tempted to leave it after continuing there six years, to go into the family of the Earl of Shelburne, now the Marquis of Lansdowne; he stipulating to give me 250 £ per annum, a house to live in, and a certainty for life in case of his death, or of my separation from him; whereas at Leeds my salary was only one hundred guineas per annum, and a house, which was not quite sufficient for the subsistence of my family, without a possibility of making a provision for them after my death.

I had been recommended to Lord Shelburne by Dr. Price, as a person qualified to be a literary companion to him. In this situation, my family being at Calne in Wiltshire, near to his Lordship's seat at Bowood, I continued seven years, spending the

summer

summer with my family, and a great part of the winter in his Lordship's house in London. My office was nominally that of *librarian*, but I had little employment as such, besides arranging his books, taking a catalogue of them, and of his manuscripts, which were numerous, and making an index to his collection of private papers. In fact I was with him as a friend, and the second year made with him the tour of Flanders, Holland, and Germany, as far as Strasburgh; and after spending a month at Paris, returned to England. This was in the year 1774.

This little excursion made me more sensible than I should otherwise have been of the benefit of foreign travel, even without the advantage of much conversation with foreigners. The very sight of new countries, new buildings, new customs, &c. and the very hearing of an unintelligible new language, gives new ideas, and tends to enlarge the mind. To me this little time was extremely pleasing, especially as I saw every thing to the greatest advantage, and without any anxiety or trouble, and had an opportunity of seeing and conversing with every person of eminence wherever we came; the political characters by his Lordship's connections,

and

and the literary ones by my own. I was soon, however, tired of Paris, and chose to spend my evenings at the hotel, in company with a few literary friends. Fortunately for me, Mr. Magellan* being at Paris, at the same time, spent most of the evenings with me ; and as I chose to return before his Lordship, he accompanied me to London, and made the journey very pleasing to me ; he being used to the country, the language, and the manners of it, which I was not. He had seen much of the world,

---

* JOHN HYACINTH De MAGELLAN a descendant of the famous Navigator Magellan, was a Portuguese Jesuit, but f r more attached to Philosophy than Christianity. He was much employed by his rich and noble correspondents abroad to procure philosophical Instruments from the Artists of Great Britain. He was a good judge of these, and being of a mechanical turn as well as a man of Science, he improved their construction in many instances. He was member of and attendant on almost all the philosophical Clubs and Meetings in London, and was generally furnished with early intelligence of philosophical discoveries from the continent. On the 17th of September 1785 he made a donation of 200 guineas to the American philosophical Society, the interest whereof was to be appropriated annually as a premium for the most useful discoveries or improvements in navigation or natural philosophy, but to the exclusion of mere natural history. He died a few years ago, leaving Mr. Nicholson and the late Dr. Crawford his Executors.   T. C.

world, and his conversation during our journey was particularly interesting to me. Indeed, in London, both before and after this time, I always found him very friendly, especially in every thing that related to my philosophical pursuits.

As I was sufficiently apprized of the fact, I did not wonder, as I otherwise should have done, to find all the philosophical persons to whom I was introduced at Paris unbelievers in christianity, and even professed Atheists. As I chose on all occasions to appear as a christian, I was told by some of them, that I was the only person they had ever met with, of whose understanding they had any opinion, who professed to believe christianity. But on interrogating them on the subject, I soon found that they had given no proper attention to it, and did not really know what christianity was. This was also the case with a great part of the company that I saw at Lord Shelburne's. But I hope that my always avowing myself to be a christian, and holding myself ready on all occasions to defend the genuine principles of it, was not without its use. Having conversed so much with unbelievers at home and abroad, I thought I should be able to combat their prejudices with some

advan-

advantage, and with this view I wrote, while I was with Lord Shelburne, the first part of my *Lettei s to a philosophical unbeliever*, in proof of the doctrines of a God and a providence, and to this I have added during my residence at Birmingham, a second part, in defence of the evidences of christianity. The first part being replied to by a person who called himself Mr. Hammon, I wrote a reply to his piece, which has hitherto remained unanswered. I am happy to find that this work of mine has done some good, and I hope that in due time it will do more. I can truly say that the greatest satisfaction I receive from the success of my philosophical pursuits, arises from the weight it may give to my attempts to defend christianity, and to free it from those corruptions which prevent its reception with philosophical and thinking persons, whose influence with the vulgar, and the unthinking, is very great.

With Lord Shelburne I saw a great variety of characters, but, of our neighbours in Wiltshire, the person I had the most frequent opportunity of seeing was Dr. Frampton, a clergyman, whose history may serve as a lesson to many. No man perhaps was ever better qualified to please in a convivial hour, or

had

had greater talents for conversation and repartee; in consequence of which, though there were several things very disgusting about him, his society was much courted, and many promises of preferment were made to him. To these, notwithstanding his knowledge of the world, and of high life, he gave too much credit; so that he spared no expence to gratify his taste and appetite, until he was universally involved in debt; and though his friends made some efforts to relieve him, he was confined a year in the county prison at a time when his bodily infirmities required the greatest indulgences; and he obtained his release but a short time before his death on condition of his living on a scanty allowance; the income of his livings (amounting to more than 400 £. per annum) being in the hands of his creditors. Such was the end of a man who kept the table in a roar.

Dr. Frampton being a high churchman, he could not at first conceal his aversion to me, and endeavoured to do me some ill offices. But being a man of letters, and despising the clergy in his neighbourhood, he became at last much attached to me; and in his distresses was satisfied, I believe, that I was one of his most sincere friends. With some great defects he
had

had some considerable virtues, and uncómmon abilities, which appeared more particularly in extempore
speaking. He always preached without notes, and
when, on some occasions, he composed his sermons,
he could, if he chose to do it, repeat the whole *verba-
tim*. He frequently extemporized in verse, in a
great variety of measures.

In Lord Shelburne's family was Lady Arabella
Denny, who is well knowh by her extensive charities. She is (for she is still living) a woman of
good understanding, and great piety. She had the
care of his Lordship's two sons until they came under
the care of Mr. Jervis, who was their tutor during
my continuance in the family. His Lordship's younger son, who died suddenly, had made astonishing
attainments both in knowledge and piety, while very
young, far beyond any thing that I had an opportunity of observing in my life.

When I went to his Lordship, I had materials for
one volume of *experiments on air*, which I soon after
published, and inscribed to him; and before I left
him I published three volumes more, and had materials for a fourth, which I published immediately
on my settling in Birmingham. He encouraged me

in

in the prosecution of my philosophical enquiries, and allowed me 40 £ per annum for expences of that kind, and was pleased to see me make experiments to entertain his guests, and especially foreigners.

Notwithstanding the attention that I gave to philosophy in this situation, I did not discontinue my other studies, especially in theology and metaphysics. Here I wrote my *Miscellaneous Observations relating to education*, and published my *Lectures on Oratory and Criticism*, which I dedicated to Lord Fitzmaurice, Lord Shelburne's eldest son. Here also I published the third and last part of my *Institutes of Natural and Revealed religion;* and having in the Preface attacked the principles of Dr. Reid, Dr. Beattie, and Dr. Oswald, with respect to their doctrine of *Common Sense*, which they made to supercede all rational inquiry into the subject of religion, I was led to consider their system in a separate work, which, though written in a manner that I do not intirely approve, has, I hope upon the whole been of service to the cause of free inquiry and truth.*

In

---

* This reply of Dr. Priestley to the Scotch Doctors, though not
written

In the preface I had expressed my belief of the doctrine of *Philosophical Necessity*, but without any design to pursue the subject, and also my great admiration of Dr. Hartley's theory of the human mind, as indeed I had taken many opportunities of doing before. This led me to publish that part of his *observations on man* which related to the doctrine of association of ideas, detached from the doctrine of vibrations, prefixing *three dissertations*, explanatory of his general system. In one of these I expressed some doubt of the immateriality of the sentient principle in man; and the outcry that was made on what I casually expressed on that subject can hardly be imagined. In all the newspapers, and most of the periodical publications, I was represented as an unbeliever in revelation, and no better than an Atheist.

This

---

written in a manner that his maturer reflection approved, compleatly set at rest the question of Common Sense as denoting the intuitive evidence of a class of moral and religious propositions capable of satisfactory proof, or of high probability from considerations *ab extra*. But Dr. Reid ought hardly to be classed with coadjutors so inferior as the Drs. Oswald and Beattie. The latter wrote something which he meant as a defence of the christian religion; but such defenders of christianity as Dr. Beattie and Soame Jenyns, are well calculated to bring it into contempt with men of reason and reflection. T. C.

This led me to give the closest attention to the subject, and the consequence was the firmest persuasion that man is wholly material, and that our only prospect of immortality is from the christian doctrine of a resurrection. I therefore digested my thoughts on the subject, and published my *Disquisitions relating to matter and spirit*, also the subjects of *Socinianism* and *necessity* being nearly connected with the doctrine of the materiality of man, I advanced several considerations from the state of opinions in antient times in favour of the former; and in a separate volume discussed more at large what related to the latter, dedicating the first volume of this work to Mr. Graham, and the second to Dr. Jebb.

It being probable that this publication would be unpopular, and might be a means of bringing odium on my patron, several attempts were made by his friends, though none by himself, to dissuade me from persisting in it. But being, as I thought, engaged in the cause of important truth, I proceeded without regard to any consequences, assuring them that this publication should not be injurious to his Lordship.

In order, however, to proceed with the greatest caution,

caution, in a business of such moment, I desired
some of my learned friends, and especially Dr.
Price, to peruse the work before it was published;
and the remarks that he made upon it led to a free
and friendly discussion of the several subjects of it,
which we afterwards published jointly; and it remains a proof of the possibility of discussing subjects
mutually considered as of the greatest importance,
with the most perfect good temper, and without the
least diminution of friendship. This work I dedicated to our common friend Mr. Lee.

In this situation I published my *Harmony of the
gospels*, on the idea of the public ministry of Jesus
having continued little more than one year, a scheme
which I first proposed in the Theological Repository; and the Bishop of Waterford having in his *Harmony* published a defence of the common hypothesis,
viz. that of its having been three years, I addressed
a *letter to him* on the subject, and to this he made a
reply in a separate work. The controversy proceeded to several publications on both sides, in the most
amicable manner, and the last *Postscript* was published jointly by us both. Though my side of the
question was without any advocates that I know of,

F   and

and had only been adopted by Mr. Mann, who seemed to have had no followers, there are few persons, I believe, who have attended to our discussion of the subject, who are not satisfied that I have sufficiently proved what I had advanced. This controversy was not finished until after my removal to Birmingham.

Reflecting on the time that I spent with Lord Shelburne, being as a guest in the family, I can truly say that I was not at all fascinated with that mode of life. Instead of looking back upon it with regret, one of the greatest subjects of my present thankfulness is the change of that situation for the one in which I am now placed; and yet I was far from being unhappy there, much less so than those who are born to such a state, and pass all their lives in it. These are generally unhappy from the want of *necessary* employment, on which account chiefly there appears to be much more happiness in the middle classes of life, who are above the fear of want, and yet have a sufficient motive for a constant exertion of their faculties; and who have always some other object besides amusement.

I used to make no scruple of maintaining, that there

there is not only most virtue, and most happiness, but even most true politeness in the middle classes of life. For in proportion as men pass more of their time in the society of their equals, they get a better established habit of governing their tempers; they attend more to the feelings of others, and are more disposed to accommodate themselves to them. On the other hand, the passions of persons in higher life, having been less controlled, are more apt to be inflamed; the idea of their rank and superiority to others seldom quits them; and though they are in the habit of concealing their feelings, and disguising their passions, it is not always so well done, but that persons of ordinary discernment may perceive what they inwardly suffer. On this account, they are really intitled to compassion, it being the almost unavoidable consequence of their education and mode of life. But when the mind is not hurt in such a situation, when a person born to affluence can lose sight of himself, and truly feel and act for others, the character is so godlike, as shews that this inequality of condition is not without its use. Like the general discipline of life, it is for the present lost

no

on the great mass, but on a few it produces what no
other state of things could do.*

The

---

* The account here given of Dr. Priestley's connection
with Lord Shelburne must be gratifying to every friend of science and literature, notwithstanding the subsequent separation.
To such persons the character of a nobleman who like Lord
Shelburne, devotes so much of his time, and so much of his income to the pursuits of knowledge, and the encouragement of those
who eminently contribute to enlighten mankind, cannot but be interesting. Had he behaved dishonourably or disrespectfully to a man
of Dr. Priestley's high station in the literary world, it would have
been an argument that science and literature were ineffectual to sof-
t n the pride of titled opulence and hereditary rank, But Ovid has
observed justly, *(ingenuas didicisse fideliter Actes, emollit mores pecsi-
nit esse feros.)*

It is right to mention an anecdote highly honourable to Lord Shelburne, on the authority of Dr. Priestley. At the conclusion of the treaty
of peace in 1783, negotiated by Lord Shelburne while he was in the
ministry, a strong opposition was expected, particularly from his
former coadjutors who soon after the death of Lord Rockingham had
seceded from Lord Shelburne's administration. It was suggested to
this nobleman, that it was customary for the minister for the time
being to let it be understood among the mutes of the ministerial
members, that they might expect the usual douceur for their votes on
such an occasion. Some light might be thrown on the nature and
quantum of this douceur, by the list of ministerial rewards distributed
at the close of each session, as stated publicly to the house of Com-

mons

The greatest part of the time that I spent with
Lord Shelburne I passed with much satisfaction,
his Lordship always behaving to me with uniform
politeness, and his guests with respect. But about
two years before I left him, I perceived evident
marks of dissatisfaction, though I never understood
the cause of it; and until that time he had been
even lavish on all occasions in expressing his satisfaction in my society to our common friends. When
I left him, I asked him whether he had any fault to
find with my conduct, and he said *none*.

At length, however, he intimated to Dr. Price, that
he wished to give me an establishment in Ireland,
where he had large property. This gave me an opportunity of acquainting him, that if he chose to dis-
solve the connexion, it should be on the terms expressed in the writings which we mutually signed
when

---

mons by the late Sir George Saville. Lord Shelburne without hesitation refused compliance; and declared that if his peace could not
obtain the unbought approbation of the house, it might take its
chance. The consequence was that although the address was carried
in the Lords by 72 to 59 it was lost in the Commons by 224 to 208.

T. C.

when it was formed, in consequence of which I should be entitled to an annuity of an hundred and fifty pounds, and then I would provide for myself, and to this he readily acceded. He told Dr. Price that he wished our separation to be amicable, and I assured him that nothing should be wanting on my part to make it truly so. Accordingly, I expected that he would receive my visits when I should be occasionally in London, but he declined them.

However, when I had been some years settled at Birmingham, he sent an especial messenger, and common friend, to engage me again in his service, having, as that friend assured me, a deep sense of the loss of Lord Ashburton (Mr. Dunning) by death, and of Colonel Barre by his becoming almost blind, and his want of some able and faithful friend, such as he had experienced in me; with other expressions more flattering than those. I did not chuse, however, on any consideration, to leave the very eligible situation in which I now am, but expressed my readiness to do him any service in my power. His Lordship's enemies have insinuated that he was not punctual in the payment of my annuity; but the contrary is true: Hitherto nothing could have been

more

more punctual, and I have no reason to suppose that it will ever be otherwise.

At Calne I had another son born to me, whom, at Lord Shelburne's request, I called Henry.

It was at the time of my leaving Lord Shelburne that I found the great value of Mr. and Mrs. Lindsey's friendship, in such a manner as I certainly had no expectation of when our acquaintance commenced; especially by their introducing me to the notice of Mrs. Rayner, one of his hearers, and most zealous friends.

Notwithstanding my allowance from Lord Shelburne was larger than that which I had at Leeds, yet my family growing up, and my expences, on this and other accounts, increasing more than in proportion, I was barely able to support my removal. But my situation being intimated to Mrs. Rayner, besides smaller sums, with which she occasionally assisted me, she gave me an hundred guineas to defray the expence of my removal, and deposited with Mrs. Lindsey, which she soon after gave up to me, four hundred guineas, and to this day has never failed giving me every year marks of her friendship. Her's is, indeed, I seriously think, one of the first christian

F 4　　　　　　characters

characters that I was ever acquainted with, having a cultivated comprehensive mind, equal to any subject of theology or metaphysics, intrepid in the cause of truth, and most rationally pious.

Spending so much of my time in London was the means of increasing my intimacy with both Mr. Lindsey and Mr. Lee, our common friend; who amidst the bustle of politics, always preserved his attachment to theology, and the cause of truth. The Sunday I always spent with Mr. Lindsey, attending the service of his chapel, and sometimes officiating for him; and with him and Mrs. Lindsey I generally spent the evening of that day at Mr. Lee's who then admitted no other company, and seldom have I enjoyed society with more relish.

My winter's residence in London was the means of improving my acquaintance with Dr. Franklin. I was seldom many days without seeing him, and being members of the same club, we constantly returned together. The difference with America breaking out at this time, our conversation was chiefly of a political nature; and I can bear witness, that he was so far from promoting, as was generally supposed, that he took every method in his power to prevent a rupture

ture

ture between the two countries. He urged so much
the doctrine of forbearance, that for some time he was
unpopular with the Americans on that account, as
too much a friend to Great Britain. His advice to
them was to bear every thing for the present, as they
were sure in time to out grow all their grievances ; as
it could not be in the power of the mother country
to oppress them long.

He dreaded the war, and often said that, if the difference should come to an open rupture, it would be
a war of *ten years*, and he should not live to see the
end of it. In reality the war lasted near eight years
but he did live to see the happy termination of it.
That the issue would be favorable to America, he
never doubted. The English, he used to say, may
take all our great towns, but that will not give them
possession of the country. The last day that he
spent in England, having given out that he should
leave London the day before, we passed together,
without any other company ; and much of the time
was employed in reading American newspapers, especially accounts of the reception which the *Boston
port bill* met with in America ; and as he read the
addresses to the inhabitants of Boston from the places

in

in the neighbourhood, the tears trickled down his cheeks.*

It is much to be lamented, that a man of Dr. Franklin's general good character, and great influence, should have been an unbeliever in christianity, and also have done so much as he did to make others unbelievers. To me, however, he acknowledged that he had not given so much attention as he ought to have done to the evidences of christianity, and desired me to recommend to him a few treatises on the subject, such as I thought most deserving of his notice, but not of great length, promising to read them, and give me his sentiments on them. Accordingly, I recommended to him Hartley's evidences of christianity in his Observations on Man, and what I had then written on the subject in my Institutes of natural and revealed religion. But the American war breaking out soon after, I do not believe that he ever found himself sufficiently at leisure for the discussion. I have kept up a correspondence with him occasionally ever since, and three

of his

* For two letters written by my father relating to Dr. Franklin and Mr Burke see appendix No. 6.

of his letters to me were, with his consent, published in his Miscellaneous Works, in quarto. The first of them, written immediately on his landing in A-merica, is very striking.

About three years before the dissolution of my connection with Lord Shelburne, Dr. Fothergill, with whom I had always lived on terms of much in-timacy, having observed, as he said, that many of my experiments had not been carried to their proper extent on account of the expence that would have attended them, proposed to me a subscription from himself and some of his fiiends, to supply me with whatever sums I should want for that purpose, and named a hundred pounds per annum. This large subscription I declined, lest the discovery of it (by the use that I should, of course, make of it) should give umbrage to Lord Shelburne, but I consented to accept of 40 £ per annum, which from that time he regularly paid me, from the contribution of him-self, Sir Theodore Jansen, Mr. Constable, and Sir George Saville.

On my leaving Lord Shelburne, which was at-tended with the loss of one half of my income, Dr. Fothergill proposed an enlargement of my allow-

ance

ance for my experiments, and likewise for my maintenance, without being under the necessity of giving my time to pupils, which I must otherwise have done. And, considering the generosity with which this voluntary offer was made by persons who could well afford it, and who thought me qualified to serve the interests of science, I thought it right to accept of it; and I preferred it to any pension from the court, offers of which were more than once made by persons who thought they could have procured one for me.

As it was my wish to do what might be in my power to shew my gratitude to my friends and benefactors that suggested the idea of writing these Memoirs, I shall subjoin a list of their names. Some of the subscriptions were made with a view to defray the expence of my experiments only; but the greater part of the subscribers were persons who were equally friends to my theological studies.

The persons who made me this regular annual allowance were Dr. Watson and his son, Mr. Wedgwood, Mr. Moseley, Mr. S. Salte, Mr. Jeffries, Mr. Radcliffe, Mr. Remington, Mr. Strutt of Derby, Mr. Shore, Mr. Reynolds of Paxton, Messrs. Galton

Galton, father and son, and the Rev. Mr. Simpson.

Besides the persons whose names appear in this list, as regular subscribers, there were other persons who, without chusing to be known as such, contributed no less to my support, and some considerably more.

My chief benefactress was Mrs. Rayner, and next to her Dr. Heberden, equally distinguished for his love of religious truth, and his zeal to promote science. Such also is the character of Mr. Tayleur of Shrewsbury, who has at different times remitted me considerable sums, chiefly to defray the expences incurred by my theological inquiries and publications.

Mr. Parker of Fleet street very generously supplied me with every instrument that I wanted in glass, particularly a capital burning lens,* sixteen inches in diameter. All his benefactions in this way would have amounted to a considerable sum. Mr. Wedgwood also, besides his annual benefaction, supplied

me

---

* Though his sight was not much worse than before during the last ten years of his life it had been much injured by his experiments with the burning Lens of which he made much use in summer time.

me with every thing that I wanted made of pottery, such as retorts, tubes, &c. which the account of my experiments will shew to have been of great use to me.

On my removal to Birmingham commenced my intimacy with Mr. William Russell, whose publie spirit, and zeal in every good cause, can hardly be exceeded. My obligations to him were various and constant, so as not to be estimated by sums of money. At his proposal I doubt not, some of the heads of the congregation made me a present of two hundred pounds, to assist me in my theological publications.

Mr. Lee shewed himself particularly my friend at the time that I left Lord Shelburne, assisting me in the difficuties with which I was then pressed, and continuing to befriend me afterwards by seasonable benefactions.   By him it was hinted to me during the administration of Lord Rockingham, with whom he had great influence, that I might have a pension from the government, to assist in defraying the expence of my experiments. Another hint of the same kind was given me in the beginning of Mr. Pitt's administration by a Bishop in whose power it was to

have

have procured it from him. But in both cases I declined the overture, wishing to preserve myself independent of every thing connected with the court, and preferring the assistance of generous and opulent individuals, lovers of science, and also lovers of liberty. Without assistance I could not have carried on my experiments at all, except on a very small scale, and under great disadvantages.

Mr. Galton, before I had any opportunity of being personally acquainted with him, had, on the death of Dr. Fothergill, taken up his subscription. His son did the same, and the friendship of the latter has added much to the happiness of my situation here.* Seldom, if ever, have I known two persons of such cultivated minds, pleasing manners, and liberal dispositions, as he and Mrs. Galton. The latter had the greatest attachment imaginable to my wife.

Mr. Salte was zealous in promoting the subscriptions to my experiments, and moreover proposed to take one of my sons as an apprentice without any fee. But my brother-in-law making the same offer, I gave it the preference : Mr. Wedgwood, who has

distin-

---

* Birmingham.

distinguished himself by his application to philosophical pursuits, as well as by his great success in the improvement of his manufactory, was very zealous to serve me, and urged me to accept of a much larger allowance than I chose.

The favours that I received from my two brothers-in-law deserve my most grateful acknowledgments. They acted the part of kind and generous relations; especially at the time when I most wanted assistance. It was in consequence of Mr. John Wilkinson's proposal, who wished to have us nearer to him, that, being undetermined where to settle, I fixed upon Birmingham, where he soon provided a house for me.

My apology for accepting of these large benefactions is, that besides the great expence of my philosophical and even my theological studies, and the education of three sons and a daughter, the reputation I had, justly or unjustly, acquired brought on me a train of expences not easy to describe, to avoid or to estimate ; so that without so much as keeping a horse (which the kindness of Mr. Russel made unnecessary) the expence of housekeeping, &c. was more than double the amount of any regular income that I had.

I

I consider my settlement at Birmingham as the happiest event in my life, being highly favorable to every object I had in view, philosophical or theological. In the former respect I had the convenience of good workmen of every kind, and the society of persons eminent for their knowledge of chemistry, particularly Mr. Watt, Mr. Keir, and Dr. Withering. These with Mr. Boulton, and Dr. Darwin, who soon left us by removing from Litchfield to Derby, Mr. Galton, and afterwards Mr. Johnson of Kenelworth and myself dined together every month, calling ourselves the *lunar society*, because the time of our meeting was near the full moon.

With respect to theology, I had the society of Mr. Hawkes, Mr. Blyth, and Mr. Scholefield, and his assistant Mr. Coates, and, while he lived Mr. Palmer, before of Macclesfield. We met and drank tea together, every fortnight, At this meeting we read all the papers that were sent for the Theological Repository, which I revived some time after my coming hither, and in general our conversation was of the same cast as that with my fellow tutors at Warrington.

Within a quarter of a year of my coming to reside

at

at Birmingham, Mr. Hawkes resigned, and I had an unanimous invitation to succeed him, as colleague with Mr. Blyth, a man of a truly christian temper. The congregation we serve is the most liberal, I believe, of any in England; and to this freedom the unwearied labours of Mr. Bourne had eminently contributed.

With this congregation I greatly improved my plan of catechizing and lecturing, and my classes have been well attended. I have also introduced the custom of expounding the scriptures as I read them, which I had never done before, but which I would earnestly recommend to all ministers. My time being much taken up with my philosophical and other studies, I agreed with the congregation to leave the business of baptizing, and visiting the sick, to Mr. Blyth, and to confine my services to the Sundays. I have been minister here between seven and eight years, without any interruption of my happiness; and for this I am sensible I am in a great measure indebted to the friendship of Mr. Russell.

Here I have never long intermitted my philosophical pursuits, and I have published two volumes of experiments, besides communications to the Royal Society.

In

In theology I have completed my friendly controversy with the Bishop of Waterford on the duration of
Christ's ministry, I have published a variety of single
sermons, which, with the addition of a few others,
I have lately collected, and published in one volume,
and I am now engaged in a controversy of great extent, and which promises to be of considerable consequence, relating to the person of Christ.

This was occasioned by my *History of the Corruptions of Christianity*, which I composed and published
presently after my settlement at Birmingham, the
first section of which being rudely attacked in the
Monthly Review,* then by Dr. Horsely, and afterwards
by Mr. Howes, and other particular opponents, I undertook to collect from the original writers the state of
opinions on the subject in the age succeeding that of
the apostles, and I have published the result of my
investigation in my *History of early opinions concerning*

---

* Written by Mr. Badcock. Mr. Badcock was originally
a dissenting minister. He came to pay his respects to my father at
Calne, at which time he agreed with him upon most subjects. He
afterwards found reason to change his opinions, or at least his conduct, connecting himself with the Clergy of the Church of England,
and became my father's bitter enemy.

ning Jesus Christ, in four volumes octavo. This work has brought me more antagonists, and I now write a pamphlet annually in defence of the unitarian doctrine against all my opponents.

My only Arian antagonist is Dr. Price, with whom the discussion of the question has proceeded with perfect amity. But no Arian has as yet appeared upon the ground to which I wish to confine the controversy, viz. the state of opinions in the primitive times, as one means of collecting what was the doctrine of the apostles, and the true sense of scripture on the subject.

Some years ago I resumed the *Theological Repository* in which I first advanced my objections to the doctrine of the miraculous conception of Jesus, and his natural fallibility and peccability. These opinions gave at first great alarm, even to my best friends; but that is now in a great measure subsided. For want of sufficient sale, I shall be obliged to discontinue this Repository for some time.

At present I thank God I can say that my prospects are better than they have ever been before, and my own health, and that of my wife, better established, and my hopes as to the dispositions and future settlement of my children satisfactory.

I shall

I shall now close this account of myself with some observations of a general nature, but chiefly an account of those circumstances for which I have more particular reason to be thankful to that good being who has brought me hitherto, and to whom I trust I habitually ascribe whatever my partial friends think the world indebted to me for,

I. Not to enlarge again on what has been mentioned already, on the fundamental blessings of a religious and liberal education, I have particular reason to be thankful for a happy temperament of body and mind, both derived from my parents. My father, grand father, and several branches of the family, were remarkably healthy, and long lived; and though my constitution has been far from robust, and was much injured by a consumptive tendency, or rather an ulcer in my lungs, the consequence of improper conduct of myself when I was at school (being often violently heated with exercise, and as often imprudently chilled by bathing, &c.) from which with great difficulty I recovered, it has been excellently adapted to that studious life which has fallen to my lot.

I have never been subject to head-achs, or any

other

other complaints that are peculiarly unfavourable
to study. I have never found myself less disposed,
or less qualified, for mental exertions of any kind at
one time of the day more than another; but all seasons have been equal to me, early or late, before dinner or after, &c. And so far have I been from suffering by my application to study, (which however
has never been so close or intense as some have imagined) that I have found my health improving from
the age of eighteen to the present time; and never
have I found myself more free from any disorder
than at present. I must, however, except a short
time preceding and following my leaving Lord Shelburne, when I laboured under a bilious complaint,
in which I was troubled with gall stones, which
sometimes gave me exquisite pain. But by confining myself to a vegetable diet, I perfectly recovered;
and I have now been so long free from the disorder
that I am under no apprehension of its return.

It has been a singular happiness to me, and a proof,
I believe, of a radically good constitution, that I have
always slept well, and have awaked with my faculties
perfectly vigorous, without any disposition to drow-
siness. Also, whenever I have been fatigued with

any

any kind of exertion, I could at any time sit down and sleep; and whatever cause of anxiety I may have had, I have almost always lost sight of it when I have got to bed; and I have generally fallen asleep as soon as I have been warm.*

I even think it an advantage to me, and am truly thankful for it, that my health received the check that it did when I was young; since a muscular habit from high health, and strong spirits, are not, I think, in general accompanied with that sensibility of mind, which is both favourable to piety, and to speculative pursuits.†

To a fundamentally good constitution of body, and the being who gave it me, I owe an even chearfulness of temper, which has had but few interruptions.

---

* My father was an early riser. He never slept more than six hours. He said he did not remember having lost a whole night's sleep but once, though when awake he often had to suffer much from pain and sickness as well as from other circumstances of a very afflictive nature.

† Though not a muscular man he went through great exertion at various times of his life with activity. He walked very firmly, and expeditiously.

G 4

ons. This I inherit from my father, who had uniformly better spirits than any man that I ever knew,
and by this means was as happy towards the close of
life, when reduced to poverty, and dependent upon
others, as in his best days; and who, I am confident,
would not have been unhappy, as I have frequently
heard him say, in a workhouse.

Though my readers will easily suppose that, in
the course of a life so full of vicissitude as mine has
been, many things must have occurred to mortify
and discompose me, nothing has ever depressed my
mind beyond a very short period. My spirits have
never failed to recover their natural level, and I have
frequently observed, and at first with some surprize,
that the most perfect satisfaction I have ever felt has
been a day or two after an event that afflicted me the
most, and without any change having taken place
in the state of things. Having found this to be the
case after many of my troubles, the persuasion that
it *would* be so, after a new cause of uneasiness, has
never failed to lessen the effect of its first impression,
and together with my firm belief of the doctrine of
necessity, (and consequently that of every thing
being ordered for the best) has contributed to that
degree

degree of composure which I have enjoyed through life, so that I have always considered myself as one of the happiest of men.

When I was a young author, (though I did not publish any thing until I was about thirty) strictures on my writings gave me some disturbance, though I believe even then less than they do most others; but after some time, things of that kind hardly affected me at all, and on this account I may be said to have been well formed for public controversy.*   But what has always made me easy in any controversy in which I have been engaged, has been my fixed reso-lution frankly to acknowledge any mistake that I might perceive I had fallen into.   That I have never been in the least backward to do this in matters of philosophy, can never be denied.

As I have not failed to attend to the phenomena of my own mind, as well as to those of other parts of nature,

---

* Though Dr. Priestley has been considered as fond of controversy and that his chief delight consisted in it, yet it is far from being true. He was more frequently the defendant than the assailant.   His con-troversies as far as it depended upon himself were carried on with temper and decency.   He was never malicious nor even sarcastic or indignant unless provoked.   T. C.

nature, I have not been insensible of some great defects, as well as some advantages, attending its constitution; having from an early period been subject to a most humbling failure of recollection, so
that I have sometimes lost all ideas of both persons
and things, that I have been conversant with. I
have so completely forgotten what I have myself
published, that in reading my own writings, what I
find in them often appears perfectly new to me, and
I have more than once made experiments the results
of which had been published by me.

I shall particularly mention one fact of this kind,
as it alarmed me much at the time, as a symptom
of all my mental powers totally failing me, until I was
relieved by the recollection of things of a similar nature having happened to me before. When I was
composing the *Dissertations* which are prefixed to my
Harmony of the-gospels, I had to ascertain something
which had been the subject of much discussion relating to the Jewish passover (I have now forgotten
what it was) and for that purpose had to consult,
and compare several writers. This I accordingly
did, and digested the result in the compass of a few
paragraphs, which I wrote in short hand. But ha-

ving

ving mislaid the paper, and my attention having been drawn off to other things, in the space of a fortnight, I did the same thing over again ; and should never have discovered that I had done it twice, if, after the second paper was transcribed for the press, I had not accidentally found the former, which I viewed with a degree of terror.

Apprized of this defect, I never fail to note down as soon as possible every thing that I wish not to forget. The same failing has led me to devise, and have recourse to, a variety of mechanical expedients to secure and arrange my thoughts, which have been of the greatest use to me in the composition of large and complex works ; and what has excited the wonder of some of my readers, would only have made them smile if they had seen me at work. But by simple and mechanical methods one man shall do that in a month, which shall cost another, of equal ability, whole years to execute. This methodical arrangement of a large work is greatly facilitated by mechanical methods, and nothing contributes more to the perspicuity of a large work, than a good arrangement of its parts.

What I have known with respect to myself has
tended

tended much to lessen both my admiration, and my
contempt, of others.   Could we have entered into
the mind of Sir Isaac Newton, and have traced all the
steps by which he produced his great works, we
might see nothing very extraordinary in the process.
And great powers with respect to some things are generally attended with great defects in others; and
these may not appear in a man's writings.   For this
reason it seldom happens but that our admiration of
philosophers and writers is lessened by a personal
knowledge of them.

As great excellencies are often balanced by great,
though not apparent, defects, so great and apparent
defects are often accompanied by great, though not
apparent, excellencies.   Thus my defect in point of
recollection, which may be owing to a want of sufficient coherence in the association of ideas formerly
impressed, may arise from a mental constitution more
favourable to new associations;   so that what I have
lost with respect to memory, may have been compensated by what is called invention, or new and original combinations of ideas.   This is a subject that
deserves attention, as well as every thing else that
relates to the affections of the mind.

Though

Though I have often composed much in a little time, it by no means follows that I could have done much in a given time. For whenever I have done much business in a short time, it has always been with the idea of having time more than sufficient to do it in ; so that I have always felt myself at ease, and I could have done nothing, as many can, if I had been hurried.

Knowing the necessity of this state of my mind to the dispatch of business, I have never put off any thing to the last moment ; and instead of doing that on the morrow which ought to be done to day, I have often blamed myself for doing to day what had better have been put off until to morrow ; precipitancy being more my fault than procrastination.

It has been a great advantage to me that I have never been under the necessity of retiring from company in order to compose any thing. Being fond of domestic life, I got a habit of writing on any subject by the parlour fire, with my wife and children about me, and occasionally talking to them, without experiencing any inconvenience from such interruptions. Nothing but reading, or speaking without interruption has been any obstruction to me. For I could not

help

help attending (as some can) when others spoke in my hearing. These are useful habits, which studious persons in general might acquire, if they would; and many persons greatly distress themselves, and others, by the idea that they can do nothing except in perfect solitude or silence.

Another great subject of my thankfulness to a good providence is my perfect freedom from any embarrassment in my circumstances, so that, without any anxiety on the subject, my supplies have always been equal to my wants; and now that my expences are increased to a degree that I had no conception of some years ago, I am a richer man than I was, and without laying myself out for the purpose. What is more, this indifference about an increase of fortune has been the means of attaining it. When I began my experiments, I expended on them all the money I could possibly raise, carried on by my ardour in philosophical investigations, and entirely regardless of consequences, except so far as never to contract any debt; and if this had been without success, my imprudence would have been manifest. But having succeeded, I was in time more than indemnified for all that I had expended.

My

My theological studies, especially those which made it necessary for me to consult the Christian Fathers, &c. have also been expensive to me. But I have found my theological friends even more liberal than my philosophical ones, and all beyond my expectations.

In reflecting on my past life I have often thought of two sayings of Jacob. When he had lost one of his sons, and thought of other things that were afflictions to him, he said, " all these things are against me," at the same time that they were in reality making for him. So the impediment in my speech, and the difficulties of my situation at Needham, I now see as much cause to be thankful for, as for the most brilliant scenes in my life.

I have also applied to myself what Jacob said on his return from Padan Aram. " With my staff I went over this Jordan, and now I am become two bands;" when I consider how little I carried with me to Needham and Nantwich, how much more I had to carry to Warrington, how much more still to Leeds, how much more than that to Calne, and then to Birmingham.

Yet, frequently as I have changed my situation,

and

and always for the better, I can truly say that I never wished for any change on my own account. I should have been contented even at Needham, if I could have been unmolested, and had bare necessaries. This freedom from anxiety was remarkable in my father, and therefore is in a manner hereditary to me ; but it has been much increased by reflection ; having frequently observed, especially with respect to christian ministers, how often it has contributed to embitter their lives, without being of any use to them. Some attention to the improvement of a man's circumstances is, no doubt, right, because no man can tell what occasion he may have for money, especially if he have children, and therefore I do not recommend my example to others. But I am thankful to that good providence which always took more care of me than I ever took of myself.

Hitherto I have had great reason to be thankful with respect to my children, as they have a prospect of enjoying a good share of health, and a sufficient capacity for performing the duties of their stations. They have also good dispositions, and as much as could be expected at their age, a sense of religion. But as I hope they will live to see this work, I say the

the less on this subject, and I hope they will consider what I say in their favour as an incitement to
exert themselves to act a christian and useful part
in life; that the care that I and their mother have
taken of their instruction may not be lost upon them,
and that they may secure a happy meeting with us
in a better world.

I esteem it a singular happiness to have lived in
an age and country, in which I have been at full liberty both to investigate, and by preaching and writing to propagate, religious truth; that though the
freedom I have used for this purpose was for some
time disadvantageous to me, it was not long so,
and that my present situation is such that I can with
the greatest openness urge whatever appears to me
to be the truth of the gospel, not only without giving
the least offence, but with the intire approbation of
those with whom I am particularly connected.

As to the dislike which I have drawn upon myself by my writings, whether that of the Calvinistic
party, in or out of the church of England, those
who rank with rational dissenters (but who have
been exceedingly offended at my carrying my inquiries farther than they wished any person to do)

or whether they be unbelievers, I am thankful that it gives less disturbance to me than it does to themselves; and that their dislike is much more than compensated by the cordial esteem and approbation of my conduct by a few, whose minds are congenial to my own, and especially that the number of such person increases.   [*Birmingham*, 1787.

A Continuation of the Memoirs, written at Northumberland in America in the beginning of the year 1795.

WHEN I wrote the preceding part of these Memoirs I was happy as must have appeared in the course of them, in the prospect of spending the remainder of my life at Birmingham, where I had every advantage for pursuing my studies, both philosophical and theological; but it pleased the sovereign disposer of all things to appoint for me other removals, and the manner in which they were brought about were more painful to me than the removals themselves. I am far, however, from questioning the wisdom or the goodness of the appointments respecting myself or others.

To resume the account of my pursuits where the former part of the Memoirs left it, I must observe that, in the prosecution of my *experiments*, I was led to maintain the doctrine of phlogiston against Mr. Lavoisier and other chemists in France, whose opinions were adopted not only by almost all the philosophers of that country, but by those in England and Scotland also. My friends, however, of the lunar society were never satisfied with the Anti-phlogistic doctrine. My experiments and observations on this subject were published in various papers in the Philosophical Transactions. At Birmingham I also published a new edition of my publications on the subject of *air*, and others connected with it, reducing the six volumes to three, which, with his consent, I dedicated to the prince of Wales.

In theology I continued my *defences of Unitarianism*, until it appeared to myself and my friends that my antagonists produced nothing to which it was of any consequence to reply. But I did not, as I had proposed, publish any address to the bishops, or to the legislature, on the subject. The former I wrote, but did not publish. I left it, however, in the hands of Mr. Belsham when I came to America, that he

H 2 might

might dispose of it as he should think proper.

The pains that I took to ascertain the state of early opinions concerning Jesus Christ, and the great mis-apprehensions I perceived in all the ecclesiastical his-torians, led me to undertake a *General History of the christian church to the fall of the Western empire*, which accordingly I wrote in two volumes octavo, and dedicated to Mr. Shore. This work I mean to continue.

At Birmingham I wrote the *second part* of my *Letters to a philosophical Unbeliever*, and dedicated the whole to Mr. Tayleur of Shrewsbury, who had afforded me most material assistance in the publica-tion of many of my theological works, without which, the sale being inconsiderable, I should not have been able to publish them at all.

Before I left Birmingham I preached a funeral sermon for my friend Dr. Price, and another for Mr. Robinson of Cambridge, who died with us on a visit to preach our annual charity school sermon. I also preached the last annual sermon to the friends of the college at Hackney. All these three sermons were published.

About two years before I left Birmingham the question

question about the *test act* was much agitated both in and out of parliament. This, however, was altogether without any concurrence of mine. I only delivered, and published, a sermon on the 5th of November 1789, recommending the most peaceable method of pursuing our object. Mr. Madan, however, the most respectable clergyman in the town, preaching and publishing a most inflammatory sermon on the subject, inveighing in the bitterest manner against the Dissenters in general, and myself in particular, I addressed a number of *familiar letters to the inhabitants of Birmingham* in our defence. This produced a reply from him, and other letters from me. All mine were written in an ironical and rather a pleasant manner, and in some of the last of them I introduced a farther reply to Mr. Burn, another clergyman in Birmingham, who had addressed to me *letters on the infallibility of the testimony of the Apostles concerning the person of Christ*, after replying to his first set of Letters, in a separate publication.

From these small pieces I was far from expecting any serious consequences. But the Dissenters in general being very obnoxious to the court, and it being imagined, though without any reason, that I

H 3   had

had been the chief promoter of the measures which gave them offence, the clergy, not only in Birmingham, but through all England, seemed to make it their business, by writing in the public papers, by preaching, and other methods, to inflame the minds of the people against me. And on occasion of the celebration of the anniversary of the French revolution on July 14th, 1791, by several of my friends, but with which I had little to do, a mob encouraged by some persons in power, first burned the meeting house in which I preached, then another meeting house in the town, and then my dwelling house, demolishing my library, apparatus, and, as far as they could, every thing belonging to me. They also burned, or much damaged, the houses of many Dissenters, chiefly my friends; the particulars of which I need not recite, as they will be found in two *Appeals* which I published on the subject written presently after the riots.

Being in some personal danger on this occasion, I went to London; and so violent was the spirit of party which then prevailed, that I believe I could hardly have been safe in any other place. There, however, I was perfectly so, though I continued to

be

be an object of troublesome attention until I left the country altogether. It shewed no small degree of courage and friendship in Mr. William Vaughan to receive me into his house, and also in Mr. Salte, with whom I spent a month at Tottenham. But it shewed more in Dr. Price's congregation at Hackney, to invite me to succeed him, which they did, though not unanimously, some time after my arrival in London.

In this situation I found myself as happy as I had been at Birmingham, and contrary to general expectation, I opened my lectures to young persons with great success, being attended by many from London; and though I lost some of the hearers, I left the congregation in a better situation than that in which I found it.

On the whole, I spent my time even more happily at Hackney than ever I had done before; having every advantage for my philosophical and theological studies, in some respect superior to what I had enjoyed at Birmingham, especially from my easy access to Mr. Lindsey, and my frequent intercourse with Mr. Belsham, professor of divinity in the New College, near which I lived. Never, on this side the

H 4
grave,

grave, do I expect to enjoy myself so much as I did by the fire side of Mr. Lindsey, conversing with him and Mrs. Lindsey on theological and other subjects, or in my frequent walks with Mr. Belsham, whose views of most important subjects were, like Mr. Lindsey's, the same with my own.

I found, however, my society much restricted with respect to my philosophical acquaintance; most of the members of the Royal Society shunning me on account of my religious or political opinions, so that I at length withdrew myself from them, and gave my reasons for so doing in the Preface to my *Observations and Experiments on the generation of air from water*, which I published at Hackney. For, with the assistance of my friends, I had in a great measure replaced my Apparatus, and had resumed my experiments, though after the loss of near two years.

Living in the neighbourhood of the New College, I voluntarily undertook to deliver the lectures to the pupils on the subject of *History and General policy*, which I had composed at Warrington, and also on *Experimental Philosophy and Chemistry*, the *Heads* of which I drew up for this purpose, and afterwards published.

published. In being useful to this Institution I found a source of considerable satisfaction to myself. Indeed, I have always had a high degree af enjoyment in lecturing to young persons, though more on theological subjects than on any other.

After the riots in Birmingham I wrote *an Appeal to the Public* on the subject, and that being replied to by the clergy of the place, I wrote a *second part*, to which, though they had pledged themselves to do it, they made no reply; so that, in fact the criminality of the magistrates, and other principal High-church men at Birmingham, in promoting the riot, remains acknowledged. Indeed, many circumstances, which have appeared since that time, shew that the friends of the court, if not the prime ministers themselves, were the favourers of that riot; having, no doubt, thought to intimidate the friends of liberty by the measure.

To my Appeal I subjoined various *Addresses**
that

* Many of these addresses have been published already. In the appendix to the present life (No: 7.) will be given an arranged list of the addresses to Dr. Priestley from various bodies of men at various times of his life ; they illustrate the following positions so honourable

that were sent to me from several descriptions of persons in England, and abroad ; and from them I will not deny that I received much satisfaction, as it appeared that the friends of liberty, civil and religious, were of opinion that I was a sufferer in that cause.   From France I received a considerable number of Addresses ; and when the present *National Convention* was called, I was invited by many of the departments to be a member of it.   But I thought

---

nourable to his character, and so necessary to a just view of it. 1st That wherever he officiated as a dissenting minister, he never quitted his situation but with the sincere regrets of those among whom he had resided, and with parting testimonies of their affectionate approbation of his conduct. 2dly. That the riots at Birmingham called forth such abundant testimonies in favour of his moral conduct and eminent usefulness; that the promoters of those riots whether in church or state can have no palliation in the eye of a discerning public for their proceedings, so far as he was the object of them.   Those only use violence in opposition to argument who have no argument to use. 3dly. That his quitting England for America, was regarded as a national loss to Great Britain, and the circumstances which induced it, a national disgrace. 4thly. That his reception in this country was as honourable as his friends had reason to expect : And his demeanour since his residence here, has been such as to gain him encreased reputation and respect, among those who knew nothing of him personally before his arrival.   T. C.

thought myself more usefully employed at home, and that I was but ill qualified for a business which required knowledge which none but a native of the country could possess; and therefore declined the honour that was proposed to me.

But no addresses gave me so much satisfaction as those from my late congregation, and especially of the young persons belonging to it, who had attended my lectures. They are a standing testimony of the zeal and fidelity with which I did my duty with respect to them, and which I value highly.

Besides congratulatory addresses, I received much pecuniary assistance from various persons, and bodies of men, which more than compensated for my pecuniary losses, though what was awarded me at the Assizes fell two thousand pounds short of them. But my brother-in-law, Mr. John Wilkinson, from whom I had not at that time any expectation, in consequence of my son's leaving his employment, was the most generous on the occasion. Without any solicitation, he immediately sent me five hundred pounds, and afterwards transferred to me ten thousand pounds which he had deposited in the French funds, and until that be productive, he allows me two hundred pounds per annum.

After

After the riots, I published my *Letters to the Swedenborgian Society*, which I had composed, and prepared for the press just before.

Mr. Wakefield living in the neighbourhood of the College, and publishing at this time his objections to *public worship*, they made a great impression on many of our young men, and in his Preface he reflected much on the character of Dr. Price. On both these accounts I thought myself called upon to reply to him, which I did in a series of *Letters to a young man*. But though he made several angry replies, I never noticed any of them. In this situation I also answered Mr. Evanson's *Observations on the dissonance of the Evangelists* in a *second set of Letters to a young man*. He also replied to me, but I was satisfied with what I had done, and did not continue the controversy.

Besides the *sermon* which I delivered on my acceptance of the invitation to the meeting at Hackney, in the preface to which I gave a detailed account of my *system of catechizing*, I published two *Fast sermons* for the years 1793 and 1794, in the latter of which I gave my ideas of antient prophecies compared with the then state of Europe, and in the preface to it I

gave

gave an account of my reasons for leaving the country. I also published a *Farewell sermon.**

But the most important of my publications in this situation were a series of *Letters to the Philosophers and Politicians of France on the subject of Religion.* I thought that the light in which I then stood in that country gave me some advantage in my attempts to enforce the evidence of natural and revealed religion. I also published a set of *sermons on the evidences of revelation*, which I first delivered by public notice, and the delivery of which was attended by great numbers. They were printed just before I left England.

As the reasons for this step in my conduct are given at large in the preface to my Fast sermon, I shall not dwell upon them here. The bigotry of the country in general made it impossible for me to place my sons in it to any advantage. William had been some time in France, and on the breaking out of the troubles in that country he had embarked for America, where his two brothers met him. My own situation,

---

* These reasons, as shewing the progress and state of his mind that induced this new æra of his life, will be inserted hereafter.

ation, if not hazardous, was become unpleasant, so that I thonght my removal would be of more service to the cause of truth than my longer stay in England. At length, therefore, with the approbation of all my friends, without exceprion, but with great reluctance on my own part, I came to that resolution; I being at a time of life in which I could not expect much satisfaction as to friends and society, comparable to that which I left, in which the resumption of my philosophical pursuits must be attended with great disadvantage, and in which success in my still more favourite pursuit, the propagation of unitarianism, was still more uncertain. It was also painful to me to leave my daughter, Mr. Finch having the greatest aversion to leave his relations and friends in England.

At the time of my leaving England my son in conjunction with Mr. Cooper, and other English emigrants, had a scheme for a large settlement for the friends of liberty in general near the head of the Susquehanna in Pennsylvania. And taking it for granted that it would be carried into effect, after landing at New-York, I went to Philadelphia, and thence came to Northumberland, a town the nearest to the proposed
ed

ed settlement, thinking to reside there until some
progress had been made in it.   The settlement was
given up; but being here, and my wife and myself
liking the place, I have determined to take up my residence here, though subject to many disadvantages.
Philadelphia was excessively expensive, and this comparatively a cheap place; and my son's, settling in the
neighbourhood, will be less exposed to temptation,
and more likely to form habits of sobriety and industry.   They will also be settled at much less expence
than in or near a large town.   We hope, after some
time, to be joined by a few of our friends from England, that a readier communication will be opened
with Philadelphia, and that the place will improve,
and become more eligible in other respects.

When I was at sea, I wrote some *observations on
the cause of the present prevalence of infidelity*, which
I published, and prefixed to a new edition of the
Letters to the Philosophers and Politicians of France.
I have also published my *Fast and Farewell sermons*,
and my *small tracts* in defence of unitarianism, also a
Continuation of those Letters, and a *third part of Let-
ters to a Philosophical Unbeliever*, in answer to *Mr.
Paine's Age of Reason.*

The

The observations on the prevalence of infidelity I have much enlarged, and intend soon to print; but I am chiefly employed on the Continuation of my History of the christian church.

Northumberland, March 24, 1795, in which I have completed the sixty second year of my age.

A

CONTINUATION

OF THE

M E M O I R S

OF

Dr. JOSEPH PRIESTLEY.

[Written by his Son Joseph Priestley.]

THUS far the narrative is from my father's manuscript, and I regret extremely, with the reader, that it falls to my lot to give an account of the latter period of his valuable life.

I entertained hopes at one time, that he would have continued it himself; and he was frequently requested to do so, by me and many of his friends in the course of the year preceding his death. He had then nearly compleated all the literary works he had in view, he had arrived at that period of life

J   when,

when, in imitation of his friend Mr. Lindsey, he had determined not to preach again in public, and beyond which he probably would not have ventured to publish any work without first subjecting it to the inspection of some judicious friend.

He was requested also, in imitation of Courayer, to add at the close of his Memoirs a summary of his religious opinions. This would have counteracted the suspicions entertained by some, that they had undergone a considerable change since his coming to America; and it was thought by his friends, that such a brief and simple statement of all that appeared to him essential to the christian belief, and the christian character, would attract the attention of many readers previously indisposed to religion altogether, from not understanding its real nature, and judging of it only from the corrupt, adulterated, and complicated state, in which it is professed in all countries called christian. Unbelievers in general have no conception of the perfect coincidence of christianity with rational philosophy, of the sublime views it affords of the divine benevolence, and how powerfully it acts to promote the pleasures and lessen the evils of the present life, at the same time that it holds out

to

to us a certain prospect of a future and endless state of enjoyment. It was suggested to him also, that as his society through life had been singularly varied and extensive, and his opportunities of attaining a general knowledge of the world, and a particular knowledge of eminent political and literary characters, very great, it would contribute much to the instruction and amusement of those into whose hands his Memoirs should fall, if they were accompanied with anecdotes of the principal characters with whom he had been acquainted. For he had a fund of anecdote which he was never backward to produce for the amusement of his friends, as occasions served for introducing it. But his relations were never sarcastic or ironical, or tended to disparage the characters of the persons spoken of, unless on subjects of manifest importance to the interests of society.

He meant to have complied with the above suggestions, but being at that time very busily employed about his Comparison, and thinking his Memoirs of little value compared with the works about which he was then engaged, he put off the completion of his narrative, until his other works should be ready for the

I 2   press.

press. Unfortunately this was too late. The work he had in hand was not compleated until the 22d January, when he was very weak and suffered greatly from his disorder, and he died on the 6th of February following:

The reader will therefore make allowance for the difference between what these Memoirs might have been, and what they now are; and particularly for the part which I venture to lay before the public as a continuation of his own account.

The reasons that induced him to quit England, and the progress of his opinions and inclinations respecting that last important æra in his life, have been but briefly stated in the preceding pages by himself. But as many may peruse these Memoirs, into whose hands his appeal to the public, occasioned by the riots at Birmingham, and his Fast sermon, in which he assigns at length his reasons for leaving his native country, are not likely to fall; I think it right to present to the readers, in his own words the history of the motives that impelled him to exchange his residence in England for one in this country.

The disgraceful riots at Birmingham were certainly the chief cause that first induced my father to
think

think of leaving England, though at the time of his writing the second part of the Appeal, in August 1792, he had not come to any determination on the subject. This appears from the following passage which as it shews the progress of his discontent, and likewise the true state of his political opinions, particularly in relation to the English form of government I shall quote.—

" In this almost universal prevalence of a spirit so extremely hostile to me and my friends, and which would be gratified by my destruction, it cannot be any matter of surprise, that a son of mine should wish to abandon a country in which his father has been used as I have been, especially when it is considered that this son was present at the riot in Birmingham, exerting himself all the dreadful night of the 14th of July, to save what he could of my most valuable property; that in consequence of this his life was in imminent danger, and another young man was nearly killed because he was mistaken for him. This would probably have been his fate, if a friend had not almost perforce kept him concealed some days, so that neither myself nor his mother knew what was become of him. I had not, how-

I 3   ever,

ever, the ambition to court the honour that has
been shewn him by the national assembly of France,
and even declined the proposal of his naturaliza-
tion.   At the most, I supposed it would have been
done without any *eclat;* and I knew nothing of its
being done in so very honourable a way until I saw
the account in the public newspapers.   To whatever country this son of mine shall choose to attach
himself, I trust that, from the good principles, and
the spirit, that he has hitherto shewn, he will discharge the duties of a good citizen."

" As to myself, I cannot be supposed to feel much
attachment to a country in which I have neither
found protection, nor redress.   But I am too old,
and my habits too fixed, to remove, as I own I should
otherwise have been disposed to do, to France, or
America.   The little that I am capable of doing must
be in England, where I shall therefore continue, as
long as it shall please the supreme Disposer of all
things to permit me*.

---

* " Since this was written, I have myself, without any solicitation
on my part, been made a citizen of France, and moreover elected a
member of the present Conventional Assembly   These, I scruple not
to avow, I consider as the greatest of honours ; though, for the reasons which are now made public, I have declined accepting tho
latter."

It might have been thought that, having written so much in defence of revelation, and of Christianity in general, more perhaps than all the clergy of the church of England now living; this defence of a *common cause* would have been received as some atonement for my demerits in writing against civil establishments of christianity, and particular doctrines. But had I been an open enemy of all religion, the animosity against me could not have been greater than it is. Neither Mr. Hume nor Mr. Gibbon was a thousandth part so obnoxious to the clergy as I am ; so little respect have my enemies for christianity itself, compared with what they have for their emoluments from it."

" As to my supposed hostility to the principles of the civil constitution of this country, there has been no pretence whatever for charging me with any thing of the kind. Besides that the very catalogue of my publications will prove that my life has been devoted to literature, and chiefly to natural philosophy and theology, which have not left me any leisure for factious politics; in the few things that I have written of a political nature, I have been an avowed advocate for our mixed government by

I 4   *King,*

King, Lords, and Commons; but because I have objected to the ecclesiastical part of it, and to particular religious tenets, I have been industriously
represented as openly seditious, and endeavouring
the overthrow of every thing that is *fixed*, the enemy
of all order, and of all government."

" Every publication which bears my name is in
favour of our present form of government. But if
I had not thought so highly of it, and had seen reason for preferring a more republican form, and
had openly advanced that opinion ; I do not know
that the proposing to free discussion a system of government different from that of England, even to
Englishmen, is any crime, according to the existing
laws of this country. It has always been thought,
at least, that our constitution authorises the free proposal, and discussion, of all theoretical principles
whatever, political ones not excepted. And though
I might now recommend a very different form of government to a people who had no previous prejudices or habits, the case is very different with respect to one that *has ;* and it is the duty of every
good citizen to maintain that government of any
country which the majority of its inhabitants ap-

prove,

prove, whether he himself should otherwise prefer it, or not."

" This, however, is all that can in reason be required of any man. To demand more would be as absurd as to oblige every man, by the law of marriage, to maintain that his particular wife was absolutely the handsomest, and best tempered woman in the world; whereas it is surely sufficient if a man behave well to his wife, and discharge the duties of a good husband."

" A very great majority of Englishmen, I am well persuaded, are friends to what are called *high maxims of government*. They would choose to have the power of the crown rather enlarged than reduced, and would rather see all the Dissenters banished than any reformation made in the church. A dread of every thing tending to *republicanism* is manifestly increased of late years, and is likely to increase still more. The very term is become one of the most opprobrious in the English language. The clergy (whose near alliance with the court, and the present royal family, after having been almost a century hostile to them, is a remarkable event in the present reign) have contributed not a little to

that

that leaning to arbitrary power in the crown which has lately been growing upon us. They preach up the doctrine of passive obedience and non-resistance with as little disguise as their ancestors did in the reign of the Stuarts, and their adulation of the king and of the minister is abject in the extreme. Both Mr. Madan's sermon and Mr. Burn's reply to my Appeal discover the same spirit; and any sentiment in favour of liberty that is at all bold and manly, such as, till of late, was deemed becoming Englishmen and the disciples of Mr. Locke, is now reprobated as seditious."

" In these circumstances, it would be nothing less than madness seriously to attempt a change in the constitution, and I hope I am not absolutely insane. I sincerely wish my countrymen, as part of the human race (though, I own, I now feel no particular attachment to them on any other ground) the undisturbed enjoyment of that form of government which they so evidently approve; and as I have no favour to ask of them, or of their governors, besides mere protection, as to a stranger, while I violate no known law, and have not this to ask for any long term, I hope it will be granted me. If not, I must,

like

like many others, in all ages and all nations, submit to whatever the supreme Being, whose eye is upon us all, and who I believe intends, and will in his own time bring about, the good of all, shall ap. point, and by their means execute." [*Appeal part II page* 109. *&c.*]

The rising disinclination which the preceding pas- sage shews had taken place in my father's mind towards a longer residence in England, became con- firmed by various circumstances, particularly the determination of his sons to emigrate to America. These, together with other reasons, that finally in- fluenced his conduct on the subject of removing to this country, are stated at large as I have before ob- served in the preface to his Fast sermon for the year 1794 and I cannot so properly give them as in his own words.

" THIS discourse, and those on the *Evidences of Divine Revelation*, which will be published about the same time, being the last of my labours in this country, I hope my friends, and the public, will in- dulge me while I give the reasons of their *being* the last, in consequence of my having at length, after

much

much hesitation, and now with reluctance, come to a resolution to leave this kingdom.

After the riots in Birmingham, it was the expectation, and evidently the wish, of many persons, that I should immediately fly to France, or America. But I had no consciousness of guilt to induce me to fly my country*. On the contrary, I came directly to London, and instantly, by means of my friend Mr. Russell, signified to the king's ministers, that I *was* there, and ready, if they thought proper,

to

---

* If, instead of flying from lawless violence, I had been flying from public justice, I could not have been pursued with more rancour, nor could my friends have been more anxious for my safety. One man, who happened to see me on horseback on one of the nights in which I escaped from Birmingham, expressed his regret that he had not taken me, expecting probably some considerable reward, as he said, it was so easy for him to have done it. My friends earnestly advised me to disguise myself as I was going to London. But all that was done in that way was taking a place for me in the mail coach, which I entered at Worcester, in another name than my own. However, the friend who had the courage to receive me in London had thought it necessary to provide a dress that should disguise me, and also a method of making my escape, in case the house should have been attacked on my account; and for some time my friends would not suffer me to appear in the streets.

to be interrogated on the subject of the riot. But no notice was taken of the message.

Ill treated as I thought I had been, not merely by the populace of Birmingham, for they were the mere tools of their superiors, but by the country in general, which evidently exulted in our sufferings, and afterwards by the representatives of the nation, who refused to inquire into the cause of them, I own I was not without deliberating upon the subject of emigration; and several flattering proposals were made me, especially from France, which was then at peace within itself, and with all the world; and I was at one time much inclined to go thither, on account of its nearness to England, the agreeableness of its climate, and my having many friends there.

But I likewise considered that, if I went thither I should have no employment of the kind to which I had been accustomed; and the season of active life not being, according to the course of nature, quite over, I wished to make as much use of it as I could. I therefore determined to continue in England, exposed as I was not only to unbounded obloquy and insult, but to every kind of outrage; and after my invitation to succeed my
friend

friend Dr· Price, I had no hesitation about it. Ac·
cordingly I took up my residence where I now
am, though so prevalent was the idea of my inse·
curity, that I was not able to take the house in
my own name; and when a friend of mine took it
in *his*, it was with much difficulty that, after some
time, the landlord was prevailed upon to transfer
the lease to me-   He expressed his apprehensions,
not only of the house that I occupied, being de·
molished, but also a capital house in which he
himself resides, at the distance of no less than
twenty miles from London, whither he supposed
the rioters would go next, merely for suffering me
to live in a house of *his*.

But even this does not give such an idea of the
danger that not only myself, but every person, and
every thing, that had the slightest connection with
me, were supposed to be in, as the following.   The
managers of one of the principal charities among the
Dissenters applied to me to preach their annual ser·
mon, and I had consented.   But the treasurer a
man of fortune, who knew nothing more of me than
my name, was so much alarmed at it, that he de·
clared he could not sleep.   I therefore, to his great
relief, declined preaching at all.

When

When it was known that I was settled where I now am, several of my friends, who lived near me, were seriously advised to remove their papers, and other most valuable effects, to some place of greater safety in London. On the 14th of July, 1792, it was taken for granted by many of the neighbours, that my house was to come down, just as at Birmingham the year before. When the Hackney association was formed, several servants in the neighbourhood actually removed their goods; and when there was some political meeting at the house of Mr. Breillat, though about two miles from my house, a woman whose daughter was servant in the house contiguous to mine, came to her mistress, to entreat that she might be out of the way; and it was not without much difficulty that she was pacified, and prevailed upon to let her continue in the house, her mistress saying that she was as safe as herself.

On several other occasions the neighbourhood has been greatly alarmed on account of my being so near them. Nor was this without apparent reason. I could name a person, and to appearance a reputable tradesman, who, in the company

of

of his friends, and in the hearing of one of my late congregation at Birmingham, but without knowing him to be such, declared that, in case of any disturbance, they would immediately come to Hackney, evidently, for the purpose of mischief. In this state of things, it is not to be wondered at, that of many servants who were recommended to me, and some that were actually hired, very few could, for a long time, be prevailed upon to live with me.

These facts not only shew how general was the idea of my particular insecurity in this country; but what is of much more consequence, and highly interesting to the country at large, an idea of the general disposition to rioting and violence that prevails in it, and that the Dissenters are the objects of it. Mr. Pitt very justly observed, in his speech on the subject of the riots at Birmingham, that it was " the effervescence of the public mind." Indeed the effervescible matter has existed in this country ever since the civil wars in the time of Charles I. and it was particularly apparent in the reign of Queen Anne. But the power of government under the former princes of the House of Hanover prevented

its

its doing any mischief. The late events shew that this power is no longer exerted as it used to be, but that, on the contrary there prevails an idea, well or ill founded, that tumultuary proceedings against Dissenters will not receive any effectual discouragement. After what has taken place with respect to Birmingham, all idea of much hazard for insulting and abusing the Dissenters is entirely vanished; whereas the disposition to injure the Catholics was effectually checked by the proceedings of the year 1780. From that time *they* have been safe, and I rejoice in it. But from the year 1791, the Dissenters have been more exposed to insult and outrage than ever.

Having fixed myself at Clapton; unhinged as I had been, and having lost the labour of several years; yet flattering myself that I should end my days here, I took a long lease of my house, and expended a considerable sum in improving it. I also determined, with the assistance of my friends, to resume my philosophical and other pursuits; and after an interruption amounting to about two years, it was with a pleasure that I cannot describe, that I entered my new laboratory, and began the most com-

K

mon preparatory processes, with a view to some original inquiries. With what success I have laboured, the public has already in some measure seen, and may see more hereafter.

But though I did not choose (notwithstanding I found myself exposed to continual insult) to leave my native country, I found it necessary to provide for my sons elsewhere. My eldest son was settled in a business, which promised to be very advantageous, at Manchester; but his partner though a man of liberality himself, informed him, on perceiving the general prevalence of the spirit which produced the riots in Birmingham, that, owing to his relationship to *me*, he was under the necessity of proposing a separation, which accordingly took place.

On this he had an invitation to join another connexion, in a business in which the spirit of party could not have much affected him; but he declined it. And after he had been present at the assizes at Warwick, he conceived such an idea of this country, that I do not believe any proposal, however advantageous, would have induced him to continue in it; so much was he affected on perceiving his father treated as I had been.

Determin-

Determining to go to America, where he had no prospect but that of being a farmer, he wished to spend a short time with a person who had greatly distinguished himself in that way, and one who from his own general principles, and his friendship for myself, would have given him the best advice and assistance in his power. He, however, declined it, and acknowledged some time after, that had it been known, as it must have been, to his landlord, that he had a son of *mine* with him, he feared he should have been turned out of his farm.

My second son who was present both at the riot, and the assizes, felt more indignation still, and willingly listened to a proposal to settle in France; and there his reception was but too flattering. However, on the breaking out of the war with this country, all mercantile prospects being suspended, he wished to go to America. There his eldest and youngest brother have joined him, and they are now looking out for a settlement, having as yet no fixed views.

The necessity I was under of sending my sons out of this country was my principal inducement to send the little property that I had out of it too; so that I had nothing in England besides my library,

K 2 apparatus

apparatus, and household goods. By this, I felt myself greatly relieved, it being of little consequence where a man already turned sixty ends his days. Whatever good or evil I have been capable of, is now chiefly done ; and I trust that the same consciousness of integrity, which has supported me hitherto, will carry me through any thing that may yet be reserved for me. Seeing, however, no great prospect of doing much good, or having much enjoyment, here, I am now preparing to follow my sons ; hoping to be of some use to them in their present unsettled state, and that Providence may yet, advancing in years as I am, find me some sphere of usefulness along with them.

As to the great odium that I have incurred, the charge of *sedition*, or my being an enemy to the constitution or peace of my country, is a mere pretence for it ; though it has been so much urged, that it is now generally believed, and all attempts to undeceive the public with respect to it avail nothing at all. The whole course of my studies, from early life, shews how little *politics* of any kind have been my object. Indeed to have written so much as I have in *theology*, and to have done so much in *expe-*
rimental

rimental philosophy, and at the same time to have had my mind occupied, as it is supposed to have been, with factious politics, I must have had faculties more than human. Let any person only cast his eye over the long list of my publications, and he will see that they relate almost wholly to theology, philosophy, or general literature.

I did, however, when I was a younger man, and before it was in my power to give much attention to philosophical pursuits, write a small anonymous political pamphlet, on the *State of Liberty in this Country*, about the time of Mr. Wilkes's election for Middlesex, which gained me the acquaintance, and I may say the friendship, of Sir George Savile, and which I had the happiness to enjoy as long as he lived.

At the request also of Dr. Franklin and Dr. Fothergill, I wrote an address to the Dissenters on the subject of the approaching rupture with America, a pamphlet which Sir George Savile, and my other friends, circulated in great numbers, and it was thought with some effect.

After this I entirely ceased to write any thing on the subject of politics, except as far as the business

K 3   of

of the *Test Act*, and of *Civil Establishments of Religion*, had a connection with politics.   And though, at the recommendation of Dr. Price, I was presently after this taken into the family of the Marquis of Landsdowne, and I entered into almost all his views, as thinking them just and liberal, I never wrote a single political pamphlet, or even a paragraph in a newspaper, all the time that I was with him, which was seven years.

I never preached a political sermon in my life; unless such as, I believe all Dissenters usually preach on the fifth of November, in favour of *civil and religious liberty*, may be said to be political. And on these occasions, I am confident, that I never advanced any sentiment but such as, until of late years, would have tended to recommend, rather than render me obnoxious, to those who direct the administration of this country. And the doctrines which I adopted when young, and which were even popular then (except with the clergy, who were at that time generally disaffected to the family on the throne) I cannot abandon, merely because the times are so changed, that they are now become unpopular, and the expression and communication of them hazardous.

                                        Farther,

Farther, though I by no means disapprove of societies for political information, such as are now
every where discountenanced, and generally sup-
pressed, I never was a member of any of them ; nor,
indeed, did I ever attend any public meeting, if I
could decently avoid it, owing to habits acquired in
studious and retired life.

From a mistake of my talents and disposition, I
was invited by many of the departments in France,
to represent them in the present National Convention, after I had been made a citizen of France, on
account of my being considered as one who had been
persecuted for my attachment to the cause of liberty
here. But though the invitation was repeated with
the most flattering importunity, I never hesitated
about declining it.

I can farther say with respect to politics, concerning which I believe every Englishman has some opinion or other (and at present, owing to the peculiar
nature of the present war, it is almost the only topic
of general conversation) that, except in company, I
hardly ever think of the subject, my reading, meditation, and writing, being almost wholly engrossed by
theology, and philosophy ; and of late, as for many

K 4   years

years before the riots in Birmingham, I have spent a very great proportion of my time, as my friends well know, in my laboratory.

If, then, my real crime has not been *sedition*, or *treason*, what has it been? For every *effect* must have some adequate *cause*, and therefore the odium that I have incurred must have been owing to something in my declared sentiments, or conduct, that has exposed me to it. In my opinion, it cannot have been any thing but my open hostility to the doctrines of the established church, and more especially to all civil establishments of religion whatever. This has brought upon me the implacable resentment of the great body of the clergy; and they have found other methods of opposing me besides *argument*, and that use of the *press* which is equally open to us all. They have also found an able ally and champion in Mr. Burke, who (without any provocation except that of answering his book on the French Revolution) has taken several opportunities of inveighing against me, in a place where he knows I cannot reply to him, and from which he also knows that his accusation will reach every corner of the country, and consequently thousands of persons who

will

will never read any writings of mine*. They have had another, and still more effectual vehicle of their abuse in what are called the *treasury newspapers*, and other popular publications.

By these and others means, the same party spirit which was the cause of the riots in Birmingham, has been increasing ever since, especially in that neighbourhood. A remarkable instance of this may be seen in a *Letter* addressed, but not sent, to me from *Mr. Foley, rector of Stourbridge*, who ac-knowledges the satisfaction that he and his brethren have received from one of the grossest and coarsest pieces of abuse of me that has yet appeared, which, as a curious specimen of the kind, I inserted in the *Appendix of my Appeal*, and in which I am repre-sented as no better than Guy Fawkes, or the devil himself. This very Christian divine recommends

to

Mr. Burke having said in the House of Commons, that " I was " made a citizen of France on account of my declared hostility to the " constitution of this country," I, in the public papers, denied the charge, and called upon him for the proofs of it. As he made no reply, I said, in the preface to my Fast Sermon of the last year, p. 9, that " it sufficiently appeared that he had neither ability to maintain " his charge, nor virtue to retract it." A year more of silence on his part having now elapsed, this is become more evident than before.

to the members of the established church to decline all commercial dealings with the Dissenters, as an effectual method of exterminating them. This method has been actually adopted in many parts of England. Also great numbers of the best farmers and artizans in England have been dismissed because they would not go to the established church. *Defoe's Shortest Way with the Dissenters** would have taught the friends of the church a more effectual method still. And yet this Mr. Foley, whom I never saw, and who could not have had any particular cause of enmity to me, had, like Mr. Madan of Birmingham, a character for liberality. What, then, have we to expect from others, when we find so much bigotry and rancour in such men as these?

Many times, by the encouragement of persons from whom better things might have been expected, I have been burned in effigy along with Mr. Paine; and numberless insulting and threatening letters have been sent to me from all parts of the kingdom.†

It

---

* A tract written in a grave ironical stile, advising to hang them all.

† In one of these I was threatened with being burned alive before a slow fire.

It is not possible for any man to have conducted himself more peaceably than I have done all the time that I have lived at Clapton, yet it has not exempted me not only from the worst suspicions, but very gross insults. A very friendly and innocent club, which I found in the place, has been considered as *Jacobin* chiefly on my account; and at one time there was cause of apprehension that I should have been brought into danger for lending one of Mr. Paine's books. But with some difficulty the neighbourhood was satisfied that I was innocent.

As nothing had been paid to me on account of damages in the riot, when I published the second part of my *Appeal* to the public on the subject, it may be proper to say, that it was paid some time in the beginning of the year 1793, with interest only from the first of January of the same year, though the injury was received in July, 1791; when equity evidently required, that it ought to have been allowed from the time of the riot, especially as, in all the cases, the allowance was far short of the loss. In my case it fell short, as I have shewn, not less than two thousand pounds. And the losses sustained by the other sufferers far exceeded mine. Public jus-

tice

tice also required that, if the forms of law, local enmity or any other cause, had prevented our receiving
full indemnification, it should have been made up to
us from the public treasury ; the great end of all civil
government being protection from violence, or an indemnification for it.   Whatever we might in equity
claim, the country owes us, and, if it be just, will
some time or other pay, and with interest.

I would farther observe, that since, in a variety of
cases, money is allowed where the injury is not of a
pecuniary nature, merely because no other compen-
sation can be given, the same should have been done
with respect to me, on account of the destruction of
my manuscripts, the interruption of my pursuits,
the loss of a pleasing and advantageous situation,
&c. &c. and had the injury been sustained by a
clergyman, he would, I doubt not, have claimed, and
been allowed, very large damages on this account.
So far, however, was there from being any idea of
the kind in *my* favour, that my counsel advised me
to make no mention of my manuscript *Lectures on
the Constitution of England*, a work about as large as
that of Blackstone (as may be seen by the syllabus
of the particular lectures, sixty-three in all, publish-

ed

ed in the first edition of my *Essay on a Course of liberal Education for civil and active Life)* because it would be taken for granted that they were of a seditious nature, and would therefore have been of disservice to me with the jury. Accordingly they were, in the account of my losses, included in the article of so much *paper*. After these losses, had I had nothing but the justice of my country to look to, I must have sunk under the burden, incapable of any farther exertions. It was the seasonable generosity of my friends that prevented this, and put it in my power, though with the unavoidable loss of near two years, to resume my former pursuits.

A farther proof of the excessive bigotry of this country is, that, though the clergy of Birmingham resenting what I advanced in the first part of my *Appeal*, replied to it, and pledged themselves to go through with the enquiry along with me, till the whole truth should be investigated, they have made no reply to the *Second Part of my Appeal*, in which I brought specific charges against themselves, and other persons by name, proving them to have been the promoters and abettors of the riot ; and yet they have as much respect shown to them as ever, and the

country

country at large pays no attention to it. Had the clergy been the injured persons, and Dissenters the rioters, unable to answer the charges brought against them, so great would have been the general indignation at their conduct, that I am persuaded it would not have been possible for them to continue in the country.

I could, if I were so disposed, give my readers many more instances of the bigotry of the clergy of the chnrch of England with respect ro me, which could not fail to excite, in generous minds, equal indignation and contempt; but I forbear.*   Had I, however, foreseen what I am now witness to, I certainly should not have made any attempt to re-place my library or apparatus, and I soon repented of having done it.   But this being done, I was willing to make some use of both before another interruption of my pursuits.   I began to philoso-phize, and make experiments, rather late in life,

being

---

* At a dinner of all the Prebendaries of a cathedral church, the conversaticn turning on the riots in Birmingham, and on a clergyman having said that if I were mounted on a pile of my publications, he would set fire to them, and burn me alive, they all declared that they would bo ready to do the same.

being near forty, for want of the necessary means
of doing any thing in this way; and my pursuits
have been much interrupted by removals (never
indeed chosen by myself, but rendered necessary
by circumstances) and my time being now short, I
hoped to have had no occasion for more than one,
and that a final, remove. But the circumstances
above mentioned have induced me, though with
great and sincere regret, to undertake another,
and to a greater distance than any that I have hitherto made.

I profess not to be unmoved by the aspect of
things exhibited in this discourse. But notwithstanding this, I should willingly have awaited my
fate in my native country, whatever it had been, if
I had not had sons in America, and if I did not think
that a field of public usefulness, which is evidently
closing upon me here, might open to more advantage there.

I own also that I am not unaffected by such unexampled punishments as those of Mr. Muir and my
friend Mr. Palmer, for offences, which, if, in the eye
of reason, they be any at all, are slight, and very in-
sufficiently proved; a measure so subversive of that
freedom

freedom of speaking and acting, which has hitherto
been the great pride of Britons.   Bur the sentence of
Mr. Winterbotham, for delivering from the pulpit
what I am persuaded he never did deliver, and
which, similar evidence might have drawn upon myself, or any other dissenting minister, who was an
object of general dislike, has something in it still
more alarming*.   But I trust that conscious in-
nocence

---

* I trust that the friends of liberty, especially among the Dissenters,
will not fail to do every thing in their power to make Mr- Winterbotham's confinement, and also the sufferings of Mr. Palmer and his com-
panions, as easy to them as possible.   Having been assisted in a season of persecution myself, I should be very ill deserving of the favours
I have received, if I was not particularly desirous of recommending
such cases as theirs to general consideration.   Here difference in religious sentiment is least of all to be attended to.   On the contrary,
let those who in this respect differ the most from Mr. Winterbotham,
which is my own case, exert themselves the most in his favour.   When
men of unquestionable integrity and piety suffer in consequence of
acting (as such persons always will do) from a principle of *conscience*,
they must command the respect even of their enemies, if they also act
from principle, though they be thereby led to proceed in an opposite
direction.

The case of men of education and reflection (and who act from the
best intentions with respect to the community) committing what only
state policy requires to be considered as *crimes*, but which are allowed

on

nocence would support me as it does him, under whatever prejudiced and violent men might *do* to me, as well as *say* of me. But I see no occasion to expose myself to danger without any prospect of doing good, or to continue any longer in a country in which I am so unjustly become the object of general dislike, and not retire to another, where I have reason to think I shall be better received. And I trust that the same good Providence which has attended me hitherto, and made me happy in my present situation, and all my former ones, will attend and bless me in what may still be before me. In all events, *The will of God be done.*

I cannot refrain from repeating again, that I

leave

---

on all hands to imply no moral turpitude, so as to render them unfit for heaven and happiness hereafter, is not to be confounded with that of common felons. There was nothing in the conduct of Louis XIV. and his ministers, that appeared so shocking, so contrary to all ideas of justice, humanity and decency, and that has contributed more to render their memory execrated, than sending such men as Mr. Marolles, and other eminent Protestants, who are now revered as saints and martyrs, to the galleys, along with the vilest miscreants. Compared with this, the punishment of death would be mercy. I trust that, the Scots in general will think these measures a disgrace to their country.

L

leave my native country with real regret, never expecting to find any where else society so suited to my disposition and habits, such friends as I have here (whose attachment has been more than a balance to all the abuse I have met with from others) and especially to replace one particular Christian friend, in whose absence I shall, for some time at least, find all the world a blank. Still less can I expect to resume my favourite pursuits, with any thing like the advantages I enjoy here. In leaving this country I also abandon a source of maintenance, which I can but ill bear to lose. I can, however truly say, that I leave it without any resentment, or ill-will. On the contrary, I sincerely wish my countrymen all happiness; and when the time for reflection (which my absence may accelerate) shall come, they will, I am confident, do me more justice. They will be convinced that every suspicion they have been led to entertain to my disadvantage has been ill founded, and that I have even some claim to their gratitude and esteem. In this case, I shall look with satisfaction to the time when, if my life be prolonged, I may visit my friends in this country; and perhaps I may, notwithstanding

my

my removal for the present, find a grave (as I believe is naturally the wish of every man) in the land that gave me birth."

On the 8th day of April 1794, my father set sail from London, and arrived at New-York on the 4th of June, where he staid about a fortnight. Many persons went to meet him upon his landing, and while he staid at New-York he received addresses from various Societies, and great attention from many of the most respectable persons in the place. From thence he proceeded to Philadelphia, where he received an address from the American Philosophical Society. Independent of the above marks of respect, he was chosen by an unanimous vote of the Trustees of the University of Philadelphia, professor of Chemistry. He was likewise invited to return and stay at New-York, and open an Unitarian place of worship, which was to have been provided for him, and also to give Lectures on Experimental Philosophy to one hundred subscribers at ten dollars each. These invitations indeed he did not receive until he had been settled some little time at Northumberland. These are sufficient proofs that the citizens of this country were not insensible to

L 2   his

his merit as a Philosopher, and that they esteemed him for the part he took in the politics of Europe. That he was not invited immediately on his arrival to preach either at New-York or Philadelphia, was not from any want of respect for his character, but because Unitarianism was in a manner unknown, and by many ignorantly supposed to have some connection with infidelity. The proper evidences of christianity, the corruptions it has suffered, the monstrous additions that have been engrafted on its primitive simplicity, and the real state of the opinions of christians in the first ages of the church, were subjects that had hardly ever been discussed in this country. The controversies that had been carried on in England had not awakened attention here, and therefore though my father was known as having suffered in consequence of his opposition to the established religion of his country, yet his particular opinions were little understood. As his religious tenets became more known, these prejudices wore away, and independent of the proposal to open a place of Unitarian worship at New-York, mentioned above, I shall have occasion to state the great reason he had to be satisfied with the testimonies of

respect

respect paid to him, by the most eminent persons in the country, not merely in his character as a Philosopher, but as a preacher of the Gospel.

About the middle of July 1794 my father left Philadelphia for Northumberland, a town situated at the confluence of the North-East and West branches of the Susquehanna, and about 130 miles North-West of Philadelphia. I, and some other English gentlemen, had projected a settlement of 300,000 acres of land, about fifty miles distant from Northumberland. The subscription was filled chiefly by persons in England. Northumberland being at that time the nearest town to the proposed settlement, my father wished to see the place, and ascertain what conveniencies it would afford should he incline either to fix there permanently, or only until the settlement should be sufficiently advanced for his accommodation; he was induced likewise to retreat, at least for the summer months, into the country, fearing the effects of the hot weather in such a city as Philadelphia. He had not, as has been erroneously reported, the least concern in the projected settlement. He was not consulted in the formation of the plan of it, nor had he come to any determina-

L 3   tion

tion to join it had it been carried into effect.

The scheme of settlement was not confined to any particular class or character of men, religious, or political. It was set on foot to be as it were a rallying point for the English, who were at that time emigrating to America in great numbers, and who it was thought, would be more happy in society of the kind they had been accustomed to, than they would be, dispersed, as they now are, through the whole of the United States. It was farther thought, that by the union of industry and capital, the wilderness would soon become cultivated and equal to any other part of the country in every thing necessary to the enjoyment of life. To promote this as much as possible, the original projectors of that scheme reserved only a few shares for themselves, for which they paid the same as those who had no trouble or expence either in forming the plan, or carrying it into execution. This they did, with a view to take away all source of jealousy, and to increase the facility of settlement, by increasing the proportion of settlers to the quantity of land to be settled. Fortunately for the original proposers, the scheme was abandoned. It might and would

have

have answered in a pecuniary point of view, as the land now sells at double and treble the price then asked for it, without the advantages which that settlement would have given rise to ; but the generality of Englishmen come to this country with such erroneous ideas, and, unless previously accustomed to a life of labour, are so ill qualified to commence cultivation in a wilderness, that the projectors would most probably have been subject to still more unfounded abuse than they have been, for their well meant endeavours to promote the interests of their countrymen.

The scheme of settlement thus failing, for reasons which it is not necessary now to state, my father, struck with the beauty of the situation of Northumberland, which is universally allowed to be equal if not superior to any in the state ; believing that, from the nature of its situation, it was likely to become a great thoroughfare, and having reason to consider it as healthy as it was pleasant, the intermittents to which it has latterly been subject being then unknown, determined to settle there. Before he came to this resolution however, he had the offer of the Professorship of Chemistry in the University of

L 4   Pennsyl-

Pennsylvania, before mentioned, which would probably have yielded him 3000 dollars per annum, there being generally about 200 students in Medicine of whom about 150 attend the Chemical Lectures; as likewise the offer of a situation as Unitarian Preacher and Lecturer in Natural Philosophy as I have likewise mentioned before. At that time he had no inducement to settle at Northumberland contrary to his inclination, as his books and apparatus were still at Philadelphia, his sons had not fixed upon any place of settlement for themselves, and neither he, nor they, had purchased a single foot of land in the town or the neighbourhood of it.

The following reasons among others induced him to prefer a country to a city life. He thought that if he undertook the duties of a professor, he should not be so much at liberty to follow his favourite pursuits as he could wish, and that the expence of living at Philadelphia or New York would counterbalance the advantages resulting from his salary; and indeed, at that time he had no occasion to attend to any pecuniary considerations, as he believed his income, calculating upon his property in the French funds (which however from circumstances not necessary

to

to be stated in this place, never produced him any
thing,) to be more than equal to his wants ; but
what had greater weight with him than any thing else
was that my mother, who had been harrassed in her
mind ever since the riots at Birmingham, thought
that by living in the country, at a distance from the
cities, she should be more likely to obtain that quiet
of which she stood so much in need.

Soon after his settlement at Northumberland, many persons, with a view that his qualifications
as an instructor of youth should not be wholly
lost to the country, concurred in a plan for the establishment of a college at Northumberland. To this
scheme several subscribed from this motive alone.
Many of the principal landholders, partly from the
above and partly from motives of interest, contributed
largely both in money and land, and there was a fair
prospect, from the liberal principles upon which it
was founded, that it would have been of very great
advantage to the country. My father was requested
to draw up a plan of the course of study he would
recommend, as well as the rules for the internal ma-
nagement of the institution, and he was appointed
President. He however declined receiving any emo-
lument.

lument, and proposed giving such lectures as he was best qualified for, *gratis;* in the same manner as he had done at Hackney, and he meant to have given to the institution the use of his library and apparatus, until the students could have been furnished with them by means of the funds of the college. In consequence of the unexpected failure of some of the principal contributors, the scheme fell through at that time, and little more was done during my father's life time than to raise the skeil of a convenient building.

I shall in this place state, though I shall anticipate, in so doing, that in the year 1803 a vacancy occurred in the University of Pennsylvania, by the death of Dr. Euen, Principal of that institution. It was intimated to my father by many of the Trustees, that in case he would accept of the appointment, there was little doubt of his obtaining it; Mr. M'Kean, the present governor of the State of Pennsylvania, being among others particularly anxious that he should accept of it. In addition to the reasons that had induced him to decline the offer of the Professorship of Chemistry were to be added the weak state of his health, which would have made the idea of his having any serious engagement to fulfil, very irksome to him ; he accordingly declined it.

He

He had frequent intimations of other proposals of a similar nature that would have been made to him, had it not become generally known, that he could not accede to them from their being inconsistent with the plan of life he had laid down for himself.

I have been thus particular in the account of his reasons for settling at Northumberland, and of the different inducements offered to him to fix elsewhere, to do away the erroneous reports respecting the former, and likewise to counteract the idea that has been so industriously circulated in England, that his abilities were undervalued, that the bigotry and prejudice he had to encounter in this country, were greater than were opposed to him in England ; that his life was in consequence rendered uncomfortable, and that if he could, he would have been glad to have returned to his native country, but was restrained by a sense of shame. Some colour was given to these reports by many of his countrymen who, from motives best known to themselves, perhaps thinking thereby to excuse the inconsistency of their own conduct, corroborated the accounts, though many of them had never seen my father in this country, and had no authority whatever for assertions which were

entirely

entirely calumnies. Some currency was also given to the statement, by the false and injurious accounts published by the Duke de Liancourt, whose book if I may judge of it by that part which treats of Pennsylvania, and of this neighbourhood in particular, is not entitled to the least credit, being false in almost every particular. This my father himself has stated in a letter addressed to him.

The writer, understanding the language of the country but very imperfectly, must necessarily have been liable to many mistakes ; nor is it to be wondered at that a man who details all the tittle tattle of every table to which he is invited, and who can basely convert the hospitable reception he meets with in a strange country, into the means of turning into ridicule those who shewed him attention and meant to serve him, should be even capable of fabricating and circulating gross and injurious falsehoods respecting individuals. I should disgrace myself, in my opinion, and still more should I disgrace the high situation which my father held in the esteem of the public, were I in this work to enter into any further consideration of his attack on my father's character, satisfied that it is beyond the reach of his falsehoods and unprovoked malevolence.

My

My father would, no doubt, have been glad to have returned to England, and have enjoyed the society of his old and much valued friends; he would have rejoiced to have been nearer the centre of the Arts and Sciences; to have been joined again to his congregation and resumed his duties as a Christian Preacher; he would have been glad at the close of life, as he expresses himself, " to have found a grave in the land that gave him birth ; " but this was impossible : and no person can read the preface to his Fast Sermon, quoted above, but must be convinced of it. Though he raised the credit of his native country by the brilliancy, the extent and the usefulness of his discoveries in different branches of science ; though during his whole life he inculcated principles of virtue and religion, which the government pretended at least to believe were necessary to the well being of the state; though in no one single act of his life had he violated any law of his country or encouraged others to do so, what was the treatment he met with in that land of boasted civilization, and at the close of the 18th Century ? It is sufficiently known, and will, as it ought to do, affect the character

racter of the nation at large. Therefore, though he could have forgotten and forgiven all that was past, though the above mentioned motives would have had great weight in inducing him to return, yet there was no reason to expect that he should meet hereafter with better treatment than he had already experienced; and in consequence of this fixed persuasion he never entertained the idea of returning to live in England. He frequently talked indeed of returning to visit his friends; but when peace took place and he could have gone with safety, so comfortably was he settled in this country, and such was his opinion of the state of things in England, that he abandoned even the idea of a temporary journey thither, altogether.

But supposing the above obstacles had not existed to his return to his native country, he had no reason to be, nor was he, dissatisfied with his reception here. Independent of the attentions paid to him upon his first arrival in this country, he continued to receive marks of respect from bodies of men, and from individuals of various opinions in religion and politics, to whom he had been all his life before an utter stranger. Little reason therefore have his countrymen to represent his reception in America as unequal

qual to his merits, or to calumniate the general character of the people here. His discoveries did not add to the credit of America as they had done to that of England, yet he was not obliged to withdraw his name from its Philosophical Society, disgusted with its illiberal treatment of himself and his friends. The Americans, comparatively speaking, had little opportunity of judging of his zeal for the real interests of religion, yet he was suffered to live in peace ; and this country has not been disgraced by the destruction of a library and apparatus uniformly dedicated to the promotion of Science, and the good of mankind. It will be said that there were not such interests to oppose in America as in England. It is true, and it proves that the Americans have done well not to create such interests, and that the placing all the religious sects upon the same footing with respect to the government of the country, has effectually secured the peace of the community, at the same time that it has essentially promoted the interests of truth and virtue.

Being now settled at Northumberland with his mind at peace, and at ease in his circumstances, he seriously applied himself to those studies which

he

he had long heen compelled to desist from, and which he had but imperfectly attended to while he resided at Hackney. It is true that he spent his time there very agreeably, in a society of highly valued friends; but he did little compared to what he effected while he was at Birmingham, or what he has done during his residence here, owing to his time being very much broken in upon at Hackney by company. To prove how much he did in this country it is only necessary to refer to the list of the publications which he presented to the world in various branches of science, in theology and general literature. Here as in England, though more at leisure than formerly, he continued to apportion his time to the various occupations in which he was engaged, and strictly adhered to a regular plan of alternate study and relaxation, from which he never materially deviated.

It was while my father was at the academy that he commenced a practice which he continued until within three or four days of his death, of keeping a diary, in which he put down the occurrences of the day; what he was employed about, where he had been, and particularly an exact account of what he

had

had been reading, mentioning the names of the authors, and the number of pages he read, which was
generally a fixed number, previously determined
upon in his own mind. He likewise noted down
any hints suggested by what he read in the course of
the day. It was his custom at the beginning of each
year to arrange the plan of study that he meant to
pursue that year, and to review the general situation
of his affairs, and at the end of the year he took an
account of the progress he had made, how far he had
executed the plan he had laid down, and whether his
situation exceeded or fell short of the expectations he
had formed.

This practice was a source of great satisfaction to
him through life. It was at first adopted as a mode
of regulating his studies, and afterwards continued
from the pleasure it gave him. The greater part of
his diaries were destroyed at the riots at Birmingham, but there are still extant those for the year
1754, 1755 and several of the subsequent years.

As it will serve to shew the regularity with which
he pursued his studies, and may possibly be instructive as well as amusing to the reader, I shall give
a specimen of the manner in which he spent a year

M   while

while he was at the academy, at Daventry, and for that purpose shall select his diary for the year 1755 when he was in his 22d year. The diary contains a particular account of what he read and wrote each day, and at different periods of the year he sums up in the following manner, the progress he had made in improvement, which I give as entered at the end of the diary.

Business done in January, February and March.

Practical.

Howe's blessedness of the righteous; Bennet's pastoral care; Norris's letters and some sermons.

Controversial.

Taylor on Atonement; Hampton's Answer; Sherlock's discourses Vol. 1; Christianity not founded in Argument; Doddridge's Answer; Warburton's divine legation; Benson on the first planting of Christianity; King's Constitution of the Primitive Church.

Classics.

Josephus, Vol. 1, from page 390 to 770; Ovid's Metamorphoses to page 139; Tacitus's History, Life of Agricola, and Manners of the Germans.

Scriptures.

Scriptures.

John the Evangelist, the Acts of the Apostles the Epistles to the Romans, Galàtians, Ephesians, 1st and 2d Corinthians, in Greek; Isaiah to the 8th chapter, in Hebrew.

Mathematics.

Maclaurin's Algebra to part 2d.

Entertaining.

Irene; Prince Arthur; Ecclesiastical characters; Dryden's fables; Peruvian tales; Voyage round the world; Oriental tales; Massey's travels; Life of Hai Ebn Yokdam; History of Abdallah.

Composition.

A Sermon on the Wisdom of God; An Oration on the means of Virtue; 1st Vol. of the Institutes of Natural and Revealed Religion.

Business done from April 1st to June 23d.

Practical.

Watts's Catechism, and discourses on Catechizing; Fenelon's spiritual works Vol. 1st and half of Vol. 2d; Saurin's Sermons a few; Thomas a Kempis Book 1st to ch. 21; Cotton Mather's life; Jenning's on preaching Christianity.

M 2

Contro-

Controversial

Towgood, Gill and Breckell on Baptism ; Le Clerc on Inspiration ; Whiston's Historical preface ; Emlyn's narrative and humble enquiry , Apostolical Constitutions ; Newton on the prophecies ; Winder's History of knowledge ; Hoadly on the Sacrament ; Lowman on the Revelation ; Moral Philosopher ; Hume's Political discourses ; Middleton's fathers of the four first centuries ; Middleton and Waterland's controversy. ————— on the Demoniacs ; Goodrich's display of Human Nature.

Classics.

Cicero's 1st. Phillippic.

Historical.

Universal History Vol. 15 and 16 and to page 488 of the 17th.

Composition.

Second Vol. of the Institutes of Natural and Revealed Religion; wrote an article on Edwards's translation of the Psalms for the review.

From June 23d to September 1.

Practical Writers.

Thomas a Kempis from Ch. 21 of Book 1st ; Hartley on Man vol. 2d. May's Prayers. Holland's Sermons.

Scriptures.

Scriptures.

From the 1st Epistle of Timothy to the Revelations, and the Gospels of Matthew, Mark, and Luke, in the Greek Testament; The books of Genesis, Exodus, and Leviticus, in the Hebrew Bible.

Classics.

Ovid from Book 9th; Demosthenes 1st Phillippic and 3 Olynthiacs; Herodotus Book 1st; Homer's Iliad, Book 1, 2, 3; Sallust.

History.

Universal History from Vol. 17 p. 488 to the end of Vol. 18. Neal's History of the Puritans 4 Volumes.

Philosophy.

The Anatomical Articles in the Universal Dictionary, several principal Agebraic ones; and all the letter A.

Composition.

12 Sermons.

Business done in September.

Practical.

Holland's Sermons, Vol. 2d; Doddridge's family Expositor Vol. 1.

M 3   *Scriptures.*

Scriptures.

John the Evangelist, in Greek.

Numbers, and to the 16th Chapter in **Deuteronomy** in Hebrew.

Classics.

Homer's Iliad, 12 books.

Mathematical.

Euclid, Lib. 1, 2, 3.

History.

Universal History, Vol. 19th.

Miscellaneous.

Mason's Student ; One of Shakespeare's plays.

Composition.

4 Sermons.

Business done in October.

Practical.

Doddridge's Expositor Vol. 2d ; Common **Prayer** Book ; Fordyce's Sermons on public Institutions.

Scriptures.

Deuteronomy from Ch. 16 to the end ; Ecclesiastes and Solomon's Song in Hebrew and Greek.

Classics.

Homer's Iliad, Book P to the end.

Mathematical.

Euclid, Lib, 4, 5, 6.

Histori-

Historical.

Universal History, Vol. 20th.

Miscellaneous.

5 Shakespeares Plays.

Composition.

3 Sermons.

Business done in November.

Practical.

Abernethy's Practical Sermons.

Scriptures.

Job, in Hebrew and the Septuagint.

Philosophy Mathematics and Chemistry.

Euclid Lib. 11 and 12 slightly; Boerhave's Theory of Chemistry a good part of Vol. 1st; Rowning's Philosophy half of Vol. 1st.

Classics.

Francis's Horace, Odes 4 books.

History.

Universal History part of Vol. 3d; Jewish Antiquities. History of the Council of Trent to page 133. Anson's voyage by Walter.

Plays.

4 of Shakespeare's plays.

Composition.

2 Sermons.

M 4

Busi-

Business done in December.

Practical.

Abernethy's Posthumous sermons Vol. 2d ;
Clarke's sermons Vol. 1st. Patric on Ecclesiastes.

Scriptures.

Psalms, in the Hebrew and Septuagint.

Philosophy.

Rowning's Philosophy part 2d and 3d.

Classics.

Francis's Horace Vol. 2 and 3.

Miscellaneous and Entertaining.

Malcolm on Music, half; 4 Shakespeare's plays.
Half of the 1st Vol. of the Rambler.
Popes Ethic Epistles, a few.

History.

Paul's Council of Trent, to page 476 ; Life of the
Duke of Marlborough.

Composition.

4 Sermons.

It will be seen by this extract from his diary, that
his studies were very varied, which, as he was always persuaded, enabled him to do so much. This
he constantly attended to through life ; his chemical
and philosophical pursuits serving as a kind of relaxation

laxation from his theological studies. His miscellaneous reading, which was at all times very extensive,
comprizing even novels and.plays, still served to increase the variety. For many years of his life, he
never spent less than two or three hours a day in
games of amusement, as cards and backgammon;
but particularly chess—at which he and my mother
played regularly three games after dinner, and as
many after supper. As his children grew up, chess
was laid aside for whist or some round game at cards,
which he enjoyed as much as any of the company.
It is hardly necessary to state that he never played
for money, even for the most trifling sum.

To all these modes of relieving the mind, he added bodily exercise. Independent of his laboratory furnishing him with a good deal, as he never employed an operator, and never allowed any one even
to light a fire, he generally lived in situations which
required his walking a good deal, as at Calne, Birmingham and Hackney. Of that exercise he was
very fond. He walked well, and his regular pace
was four miles an hour. In situations where
the necessity of walking was not imposed upon
him, he worked in his garden as at Calne,
                                                      when

when he had not occasion to go to Bowood; at Northumberland in America, he was particularly attached to this exercise.

But what principally enabled him to do so much was regularity, for it does-not appear that at any period of his life he spent more than six or eight hours per day in business that required much mental exertion. I find in the same diary, which I have quoted from above, that he laid down the following daily arrangement of time for a minister's studies: Studying the Scriptures 1 hour. Practical writers 1-2 an hour. Philosophy and History 2 hours. Classics 1-2 an hour. Composition 1 hour—in all 5 hours. He adds below "All which may be " conveniently dispatched before dinner, which leaves " the afternoon for visiting and company, and the " evening for exceeding in any article if there be " occasion. Six hours not too much, nor seven."

It appears by his diary that he followed this plan at that period of his life. He generally walked out in the afternoon or spent it in company. At that time there was a society or club that assembled twice a week, at which the members debated questions, or took it in turn to deliver orations, or read essays

says of their own composition. When not attending these meetings, he most generally appears to have spent the evening in company with some of the students in their chambers.

It was by the regularity and variety of his studies, more than by intenseness of application, that he performed so much more than even studious men generally do. At the time he was engaged about the most important works, and when he was not busily employed in making experiments, he always had leisure for company, of which he was fond. He never appeared hurried or behind hand. He however never carried his complaisance so far as to neglect the daily task he had imposed upon himself; but as he was uniformly an early riser, and dispatched his more serious pursuits in the morning, it rarely happened but that he could accomplish the labours assigned for the day, without having occasion to withdraw from visitors at home, or society abroad, or giving reason to suppose that the company of others was a restraint upon his pursuits.

This habit of regularity, extended itself to every thing that he read, and every thing he did that was susceptible of it. He never read a book with-

without determing in his own mind when he would finish it. Had he a work to transcribe, he would fix a time for its completion. This habit increased upon him as he grew in years, and his diary was kept upon the plan I have before described, till within a few days of his death.

To the regularity and variety of his studies, must he added a considerable degree of Mechanical contrivance, which greatly facilitated the execution of many of his compositions. It was however most apparent in his laboratory, and displayed in the simplicity and neatness of his apparatus, which was the great cause of the accuracy of his experiments, and of the fair character which he acquired as an experimental chemist. This was the result in the first instance of a necessary attention to œconomy in all his pursuits, and was afterwards continued from choice, when the necessity no longer existed. I return from this digression which I thought necessary to give the reader a general view of my father's occupations, and his manner of spending his time, to the circumstances attending the remaining years of his life.

At his first settling at Northumberland, there was
no

no house to be procured that would furnish him with the conveniencies of a library and laboratory in addition to the room necessary for a family. Hence in the beginning of the year 1795, being then fixed in his determination to move no more, he resolved upon building a house convenient for his pursuits. During the time the house was building, he had no convenience for making experiments more than a common room afforded, and he was thereby prevented from doing much in this way. Still, he ascertained several facts of importance in the year 1795 on the Analysis of Atmospheric Air, and also some in continuation of those on the generation of air from water.

He had however leisure and opportunity for his other studies and in 1795 he published observations on the increase of infidelity and he continued his Church History from the fall of the Western Empire to the reformation.

In the spring of 1796 he spent three months at Philadelphia and delivered there a set of discourses on the Evidences of Revelation, which he composed with a view to counteract the effect produced by the writings of unbelievers, which, as might be expected,

was

was very great in a country where rational opinions in religion were but little known, and where the evidences of revelation had been but little attended to. It was a source of great satisfaction to him, and what he had little previous reason to expect, that his lectures were attended by very crowded audiences, including most of the members of the congress of the United States at that time assembled at Philadelphia, and of the executive officers of the government. These discourses which, in a regular and connected series, placed Christianity, and the evidences of its truth, in a more clear and satisfactory point of view than it had been usually considered in this country, attracted much attention, and created an interest in the subject which there is reason to believe has produced lasting effects. My father received assurances from many of the most respectable persons in the country, that they viewed the subject in a totally different light from what they had before done, and that could they attend places of worship, where such rational doctrines were inculcated, they should do it with satisfaction.

As my father had through life considered the office of a Christian minister as the most useful and honourable of any, and had always derived the greatest

satisfac-

satisfaction from fulfilling its duties, particularly from catechizing young persons, the greatest source of uneasiness therefore to him at Northumberland was, that there was no sufficient opportunity of being useful in that way. Though he was uniformly treated with kindness and respect by the people of the place, yet their sentiments in religion were so different from his own, and the nature and tendency of his opinions were so little understood, that the establishment of a place of unitarian worship perfectly free from any calvinistic or Arian tenet, was next to impossible. All therefore that he could do in that way was, for the two or three first years, to read a service either at his own or at my house, at which a few (perhaps a dozen) English persons were usually present, and in time, as their numbers increased he made use of a school room near his house, where from twenty to thirty regularly attended, and among them some of the inhabitants of the place, who by degrees began to divest themselves of their prejudices with respect to his opinions. However small the number of persons attending, he administered the Lord's supper, a rite upon which he always laid particular stress.

In the Autumn of 1795 he had the misfortune to lose

lose his youngest son, of whom being much younger than any of his other children, and having entertained the hopes of his succeeding him in his Theological and Philosophical pursuits he was remarkably fond.   He felt this misfortune the more severely
as it was the first of the kind he had experienced,
and particularly as it had a visible effect upon my
mother's health and spirits.   He was however so constantly in the habit of viewing the hand of God in
all things, and of considering every occurrence as
leading to good, that his mind soon recovered its accustomed serenity, and his journey to Philadelphia
mentioned above and the success which attended his
first exertions in the cause of, what he deemed, pure
and genuine christianity, led him to look forward
with cheerfulness to the future, and gave him an e-
nergy in his pursuits, which was never exceeded in
any part of his life.   It was the same habit of viewing God as the author of all events, and producing good out of seeming evil, that enabled him to
support himself so well under the greatest affliction
that could possibly have befallen him, viz. the loss
of his wife, my mother; who through life had been
truly a help meet for him; supporting him under
                                                                all

all his trials and sufferings with a constancy and perseverance truly praise worthy, and who as he himself, in noting the event in his diary, justly observes,
" was of a noble and generous mind and cared much
for others and little for herself through life."

In the period between the above very afflicting
events, though his conveniences for experimenting
were not increased, owing to his house, and particularly his laboratory not being finished, he wrote a
small treatise in defence of the doctrine of Phlogiston, addressed to the Philosophers in France.   He
likewise composed a second set of discourses of a
similar kind to those delivered in Philadelphia the
preceding winter.   He preached and printed a sermon in defence of Unitarianism, and printed the first
set of discourses , he compleated his Church History ; he made additional observations on the increase
of infidelity chiefly in answer to Mr. Volney ; and
drew up an Outline of all the Evidences in favour of
Revelation.

In the spring of 1797 he again spent two or three
months in Philadelphia, and delivered a second set
of discourses, but partly from the novelty of the thing
being done away, partly from the prejudices that be-

N   gan

gan to be excited against him on account of his sup-posed political opinions, (for high-toned politics began then to prevail in the fashionable circles) and partly owing to the discourses not being so well adapted for a public audience, though necessary to set the comparative excellence of Christianity in its true light, they were but thinly attended in comparison to his former set. This induced him to give up the idea of preaching any more regular sets of discourses. He however printed them, as likewise a sermon he preached in favour of the Emigrants. He also composed at this time a third and enlarged edition of his Observations on the increase of infidelity, a controversy with Mr. Volney, a tract on the Knowledge of a Future state among the Hebrews, which, with the works he composed the year before, he printed as he found means and opportunity. He revised his Church History, began his Notes on the Scriptures, and his Comparison of the Institutions of Moses with those of the Hindoos.

Towards the end of 1797 and not before, his library and laboratory were finished. None but men devoted to literature can imagine the pleasure he derived from being able to renew his experiments with

every

every possible convenience, and from having his books once more arranged. His house was situated in a garden, commanding a prospect equal, if not superior, to any on the river Susquehanna, so justly celebrated for the picturesque views its banks afford. It was a singularly fortunate circumstance that he found at Northumberland several excellent workmen in metals, who could repair his instruments, make all the new articles he wanted in the course of his experimenting, as well as, he used to say, if not in some respects better than, he could have got them done in Birmingham; and in the society of Mr. Frederick Antis, the brother of Mr. Antis in England, and uncle of Mr. Latrobe the engineer, he derived great satisfaction. Mr. Antis was a man of mild and amiable manners, he possessed a very good knowledge of Mechanics the result of his own observation and reflection, and a fund of knowledge of many things which my father frequently found useful to resort to. The situation of Northumberland became abundantly more convenient than it was when he first came to the place. From there being no regular public post, there was now established a post twice a week to Philadelphia, and answers could be

receiv-

received to letters within a week, and the communication so much increased between the two places, that the price of the carriage of goods was reduced from 11s.-3d. to 6s. per Cwt. the distance being 132 miles.

Thus conveniently situated, he resumed the same kind of life he led at Birmingham, experimenting the greater part of the day, the result of which he published in the Medical repository of New-York. Having compleated his Church History, he finished his Comparison of the Institutions of Moses with those of the Hindoos. He likewise proceeded as far as Leviticus in the design he had formed of writing Notes on all the books of Scripture, and made some remarks on the origin of all religions by Dupuis, but the greater part of the time that he spent in theology this year, was employed in recomposing the Notes on the New-Testament, which were destroyed at the riots.

In the course of the year 1799, he finished his Notes on all the books of Scripture, he published his Comparison of the Institutions of Moses with those of the Hindoos, he likewise printed his Defence of the doctrine of Phlogiston above mentioned, and the greater

greater part of each day in the summer was employ-ed in making the additional experiments he had pro-jected.

It was in the year 1799 during Mr. Adams's administration, that my father had occasion to write any thing on the subject of politics in this country. It is well known to all his friends, that politics were always a subject of secondary importance with him. He however took part occasionally in the conversa-tions on that subject; which every person has a right to do, and which, about the time my father left England, no person could avoid doing, as the subject engrossed so large a part of the conversation in al-most every company. He always argued on the side of liberty. He was however in favour only of those changes that could be brought about by fair argument, and his speculations on the subject of British politics did not go further than a reform in Parliament, and no way tended, in his opinion, to af-fect the form of government, or the constitution of the kingdom, as vested in Kings, Lords and Com-mons. He used frequently to say, and it was said to him, that though he was an Unitarian in Religion he was in that country a Trinitarian in politics.

When

When he came to America, he found reason to change his opinions, and he became a decided friend to the general principles and practice of a compleatly representative government, founded upon universal suffrage, and excluding hereditary privileges, as it exists in this country. This change was naturally produced by observing the ease and happiness with which the people lived, and the unexampled prosperity of the country, of which no European, unless he has resided in it some time, and has observed the interior part of it, can be a competent judge. But with respect to England, he still remained anxious for its peace and prosperity, and though he had been so hardly used, and though he considered the administration of the country, if not instigating at least conniving at the riots, no resentment existed in his breast against the nation. In his feelings he was still an Englishman. Though he might speculatively consider that the mass of evil and misery had arisen to such a height in England, and in other European countries, that there was no longer any hope of a peaceable and gradual reform, yet, considering at the same time that the great body of the people, like the Negroes in the West-Indies,

were

were unprepared for the enjoyment of liberty in its full extent, and contemplating the evils necessarily attendant upon a violent change, he dreaded a revolution.

With respect to America he had never interfered publicly in politics, and never wrote an article that could be considered in that light in any respect, except one published in a newspaper called the *Aurora*, signed a *Quaker in Politics*, published on the 26th and 27th of February, 1798, and entitled Maxims of Political Arithmetic,* and so little did he interest himself in the politics of this country, that he seldom if ever perused the debates in Congress, nor was he much acquainted with any of the leading political characters except three or four, and with these he never corresponded but with Mr. Adams prior to his being chosen president, and Mr. Jefferson. He never was naturalized, nor did he take part directly or indirectly in any election. He persevered in the same sentiments even when he was under reasonable apprehension that he should be banished

* See Appendix, No. IV.

N 4

nished as an Alien : and though he advised his sons to be naturalized, saying it was what was daily done by persons who could not be suspected of wishing any ill to their native country, yet he would not ; but said, that as he had been born and had lived an Englishman, he would die one let what might be the consequence.

About the year 1799, the friends of liberty in A-merica were greatly alarmed by the advancement of principles disgraceful to America, and by a practice less liberal in many respects than under the monar-chical form of the British government. Nothing else was the subject of conversation and my father who though never active in politics, at the same time never concealed his sentiments, uttered them freely, in conversation, and they were of course opposed to the proceedings of the administration at the time. Added to this Mr. Thomas Cooper formerly of Manchester, and who at that time had undertaken for a short period, at the request of the printer, to edit a newspaper then printed at Northumberland, had published some very severe strictures on the conduct of the administration, which were soon af-ter published in a pamphlet, under the title of Poli-tical Essays.

By

By many my father might be ignorantly supposed as the prompter on the occasion, as Mr. Cooper lived at that time with my father, and by those who knew better, it was made the ostensible ground of objection to my father, to conceal the real one. In truth he saw none of the essays until they were printed, nor was he consulted by Mr. Cooper upon any part of them. The consequence was, that all the bigotry and party zeal of that violent period was employed to injure him, and misrepresent his words and actions. He was represented as intriguing for offices for himself and his friend, and as an enemy to the government which they said protected him, while men who were themselves but newly naturalized, or the immediate descendants of foreigners, bestowed upon him the epithet of Alien, an epithet then used by the government party as a term of reproach, though the country was principally indebted to the capital, industry and enterprize of foreigners for the many improvements then carrying on. Such was the effect of all these slanderous reports, and such was the character of the administration, that it was intimated to my father, from Mr. Adams himself, that he wished he would abstain from saying

any

any thing on politics, lest he should get into difficulty. The Alien law which was passed under that administration, was at that time in operation, and a man without being convicted of, or even positively charged with, any offence, might have been sent out of the country at a moment's warning, not only without a trial, but without the right of remonstrance. It was likewise hinted to my father as he has himself stated, that he was one of the persons contemplated when the law was passed, so little did they know of his real character and disposition. This occasioned my father to write a set of letters to the inhabitants of Northumberland; in which he expressed his sentiments fully on all the political questions at that time under discussion. They had the effect of removing the unfavourable impressions that had been made on the minds of the liberal and candid, and procured him many friends. Fortunately however the violent measures then adopted produced a compleat change in the minds of the people, and in consequence of it in the representation, proving by the peaceableness of it, the excellence of this form of government, and proving also that my father's sentiments, as well as Mr. Cooper's, were approved of by nine tenths of the people of the United States.

It

It is but justice however to mention that in the above remarks which have been made to represent my father's political character in its true light, and to account for his writing on the subject of politics, I do not mean to reflect on all the federalists. and that though my father considered them all as in error, yet he acknowledged himself indebted to many of that party for the most sincere marks of friendship which he had received in this country, and that not only from his opponents in politics, but likewise from many of the principal clergymen of various denominations in Philadelphia, and particularly during his severe illness in that city, when party spirit was at the highest, it being at the time of Mr. Jefferson's first election to the presidency.

As my father has given an account of those friends to whose kindness and generosity he was principally indebted from the commencement of his literary career, to the time of his coming to America, I think it my duty to follow his example, and to make on his part those acknowledgements which had he lived, he would have taken pleasure in making himself. To the Revd. Theophilus Lindsey, independent of the many marks of the most sincere friendship, which he was

constant-

constantly receiving, he was occasionally indebted for pecuniary assistance at times when it was most wanting. Independent of 50 £. per annum, which Mrs. Elizabeth Rayner allowed him from the time he left England, she left him by her will £.2000 in the 4 per cents. Mr. Michael Dodson who is well known as the translator of Isaiah left him £.500, and Mr. Samuel Saite left him 100 £. The Duke of Grafton remitted him aunually 40 £. Therefore though his expences were far greater than he expected, and though his house cost him double the sum he had contemplated, the generosity of his friends made him perfectly easy in his mind with respect to pecuniary affairs; and by freeing him from all care and anxiety on this head contributed greatly to his happiness, and to his successful endeavours in the cause of truth. Besides these instances of friendly attention, the different branches of his family have been, in various ways, benefited, in consequence of the respect paid to my father's character, and the affectionate regard shewn by his friends to all who were connected with him.

But what gave my father most real pleasure was the subscription, set on foot by his friends in England,

to

to enable him to print his Church History, and his Notes on all the Books of Scripture. The whole was done without his knowledge, and the first information he received on the subject was, that there was a sum raised sufficient to cover the whole expence.

About the time he died, some of his friends in England understood that he was likely to suffer a loss in point of income of £. 200 per annum. Without any solicitation, about forty of them raised the sum of £. 450, which was meant to have been continued annually while he lived. He did not live to know of this kind exertion in his favour. It is my duty however to record this instance of generosity, and I do it with pleasure and with gratitude. It likewise proves that though my father by the fearless avowal of his opinions created many enemies, yet that the honesty and independence of his conduct procured him many friends.

The first years subscription has been transmitted to America, to defray the expence of publishing his posthumous works.

In the year 1800 he was chiefly employed in experiments, and writing an account of them for various publica-

publications. In this year also he published his treatise in defence of Phlogiston, he revised his Church History, the two first volumes of which are now reprinted with considerable additions, and he added to and improved his Notes on the Scriptures.

He spent some time in the spring of 1801 in Philadelphia, during his stay there he had a violent attack of fever which weakened him exceedingly, and from the effects of which he never perfectly recovered. Added to this the fever and ague prevailed at Northumberland and the neighbourhood, for the first time since his settlement at the place. He had two or three attacks of this disorder; which though they were not very severe, as he had never more than three fits at a time, retarded his recovery very much. He perceived the effect of his illness in the diminution of his strength, and his not being able to take as much exercise as he used to do. His spirits however were good, and he was very assiduous in making experiments, chiefly on the pile of Volta, the result of which he sent an account of to Nicholson's Journal and the Medical Repository.

In 1802 he began to print his Church History, in consequence of the subscription raised by his friends

in

in England as before stated. Besides printing three volumes of that work, he wrote and printed a treatise on Baptism, chiefly in answer to the observations of Mr. Robinson on the subject. He likewise made some experiments, and replied to some remarks of Mr. Cruikshank in defence of the Antiphlogistic theory.

I am now to describe the last scene of his life, which deserves the reader's most serious consideration, as it shews the powerful effect of his religious principles. They made him, not resigned to quit a world in which he no longer had any delight, and in which no hope of future enjoyment presented itself, but chearful in the certainty of approaching dissolution, and under circumstances that would by the world in general have been considered as highly enviable. They led him to consider death as the labourer does sleep at night as being necessary to renew his mental and corporeal powers, and fit him for a future state of activity and happiness. For though since his illness in Philadelphia in 1801 he had never recovered his former good state of health, yet he had never been confined to his bed a whole day by sickness in America until within two days of his

death

death, and was never incapacitated for any pursuit
that he had been accustomed to. He took great
delight in his garden, and in viewing the little
improvements going forward in and about the
town. The rapidly increasing prosperity of the
country, whether as it regarded its agriculture, manufactures, and commerce, or the increasing taste for
science and literature, were all of them to him a
source of the purest pleasure. For the last four
years of his life he lived under an administration,
the principles and practice of which he perfectly approved, and with Mr. Jefferson, the head of that
administration, he frequently corresponded, and they
had for each other a mutual regard and esteem. He
enjoyed the esteem of the wisest and best men in
the country, particularly at Philadelphia, where his
religion and his politics did not prevent his being
kindly and cheerfully received by great numbers of
opposite opinions in both, who thus paid homage to
his knowledge and virtue. At home he was beloved; and besides the advantages of an excellent
library, to which he was continually making additions, and of a laboratory that was amply provided
with every thing necessary for an experimental chemist,

mist, he was perfectly freed, as he had happily been through life, in consequence of my mother's ability and attention, from any attention to worldly concerns; considering himself, as he used to express himself, merely as a lodger, having all his time to devote to his theological and philosophical pursuits. He had the satisfaction of witnessing the gradual spread of his religious opinions, and the fullest conviction that he should prevail over his opponents in chemistry. He looked forward with the greatest pleasure to future exertions in both these fields, and had within the last month or six weeks been projecting many improvements in his apparatus, which he meant to make use of upon the return of warm weather in the spring. Notwithstanding, therefore, the many trials he underwent in this country, he had still great sources of happiness left, unalloyed by any apprehension of any material defect in any of his senses, or any abatement of the vigour of his mind. Consistent with the above was his declaration that, excepting the want of the society of Mr. L. Mr. B. and two or three other particular friends, which however was made up to him, in some, though in a small degree by their regular correspondence, he

O   had

had never upon the whole spent any part of his life more happily, nor, he believed, more usefully.

The first part of his illness, independent of his general weakness, the result of his illness in Philadelphia in 1801, was a constant indigestion, and a difficulty of swallowing meat or any kind of solid food unless previously reduced by mastication to a perfect pulp. This gradually increased upon him till he could swallow liquids but very slowly, and led him to suspect, which he did to the last, that there must be some stoppage in the œsophagus. Latterly he lived almost entirely upon tea, chocolate, soups, sago, custard puddings, and the like. During all this time of general and increasing debility, he was busily employed in printing his Church History, and the first volume of the Notes on Scripture ; and in making new and original experiments, an account of which he sent to the American Philosophical Society in two numbers, one in answer to Dr. Darwin's observations on Spontaneous generation, and the other on the unexpected conversion of a quantity of the marine acid into the nitrous. During this period, likewise, he wrote his pamphlet of Jesus and Socrates compared, and re-printed his

Essay

Essay on Phlogiston.   He would not suffer any one
to do for him what he had been accustomed to do
himself; nor did he alter his former mode of life in
any respect, excepting that he no longer worked in
his garden, and that he read more books of a miscellaneous nature than he had been used to do when
he could work more in his laboratory, which had
always served him as a relaxation from his other
studies.

From about the beginning of November 1803,
to the middle of January 1804, his complaint grew
more serious.   He was once incapable of swallowing
any thing for near thirty hours; and there being
some symtoms of inflammation at his stomach,
blisters were applied, which afforded him relief; and
by very great attention to his diet, riding out in a
chair when the weather would permit, and living
chiefly on the soft parts of oysters, he seemed if not
gaining ground, at least not getting worse; and we
had reason to hope that if he held out until spring as
he was, the same attention to his diet with more exercise, which it was impossible for him to take on account of the cold weather, would restore him to
health.   He, however, considered his life as very

precari-

precarious, and used to tell the physician who attended him, that if he could but patch him up for six months longer he should be perfectly satisfied, as he should in that time be able to complete printing his works. The swelling of his feet, an alarming symptom of general debility, began about this time.

To give some idea of the exertions he made even at this time, it is only necessary for me to say, that besides his miscellaneous reading, which was at all times very great, he read through all the works quoted in his comparison of the different systems of the Grecian Philosophers with christianity, composed that work, and transcribed the whole of it in less than three months. He took the precaution of transcribing one day in long hand what he had composed the day before in short hand, that he might by that means leave the work complete as far as it went, should he not live to complete the whole. During this period he composed in a day his second reply to Dr. Linn.

About this time he ceased performing divine service, which he said he had never before known himself incapable of performing, notwithstanding he had

been

been a preacher so many years. He likewise now suffered me to rake his fire, rub his feet with a flesh-brush, and occasionally help him to bed. In the mornings likewise he had his fire made for him, which he always used to do himself, and generally before any of the family was stirring.

In the last fortnight in January he was troubled with alarming fits of indigestion; his legs swelled nearly to his knees, and his weakness increased very much. I wrote for him, while he dictated, the concluding section of his New Comparison, and the Preface and Dedication. The finishing this work was a source of great satisfaction to him, as he considered it as a work of as much consequence as any he had ever undertaken. The first alarming symptom of approaching dissolution was his being unable to speak to me upon my entering his room on Tuesday morning the 31st of January. In his Diary I find he stated his situation as follows: " Ill all day— Not able to speak for near three hours." When he was able to speak he told me he had slept well, as he uniformly had done through the whole of his illness; so that he never would suffer me, though I frequently requested he would do it, to sleep in the same room with him ; that he felt as wall as possible ; that he got up and shaved himself, which he never omitted doing every morning till within two days of his death ; that he went to his laboratory, and then found his weakness very great ; that he got back with difficulty ; that just afterward his grand-daughter, a child of about six or seven years old, came to him to claim the fulfilment of a promise he had made her the evening before, to give her a fivepenny bit. He gave her the money, and was going to speak to her, but found himself unable. He informed me of this, speaking very slowly a word at a time ; and added, that he had never felt more pleasantly in his whole life than he did during the time he was unable to speak. After he had taken his medicine, which was bark and laudanum, and drank a bason of strong mutton broth, he recovered surprizingly, and talked with cheerfulness to all who called upon him, but as though he was fully sensible that he had not long to live. He consented for the first time that I should sleep in the room with him.

On Wednesday, February 1, he writes, " I was at times much better in the morning: capable of some business: continued better all day." He

spake

spake this morning as strong as usual, and took in the course of the day a good deal of nourishment with pleasure. He said, that he felt a return of strength, and with it there was a duty to perform. He read a good deal in Newcome's Translation of the New Testament, and Stevens's History of the War. In the afternoon he gave me some directions how to proceed with the printing his work in case he should die. He gave me directions to stop the printing of the second volume, and to begin upon the third, that he might see how it was begun, and that it might serve as a pattern to me to proceed by.

On Thursday, the 2d, he wrote thus for the last time in his Diary: " Much worse: incapable of business: Mr. Kennedy came to receive instructions about printing in case of my death." He sat up, however, a great part of the day, was cheerful, and gave Mr. Cooper and myself some directions, with the same composure as though he had only been about to leave home for a short time. Though it was fatiguing to him to talk, he read a good deal in the works above mentioned.

On Friday he was much better. He sat up a

good

good part of the day reading Newcome; Dr. Disney's Translation of the Psalms; and some chapters
in the Greek Testament, which was his daily practice.
He corrected a proof-sheet of the Notes on Isaiah.
When he went to bed he was not so well: he had
an idea he should not live another day. At prayer-
time he wished to have the children kneel by his
bedside, saying, it gave him great pleasure to see the
little things kneel; and, thinking he possibly might
not see them again, he gave them his blessing.

On Saturday, the 4th, my father got up for
about an hour while his bed was made. He said he
felt more comfortable in bed than up. He read a
good deal, and looked over the first sheet of the third
volume of the Notes, that he might see how we were
likely to go on with it; and having examined the
Greek and Hebrew quotations, and finding them
right, he said he was satisfied we should finish the
work very well. In the course of the day, he expressed his gratitude in being permitted to die quiet-
ly in his family, without pain, with every convenience
and comfort he could wish for. He dwelt upon the
peculiarly happy situation in which it had pleased the
Divine Being to place him in life; and the great advantage

vantage he had enjoyed in the acquaintance and friendship of some of the best and wisest men in the age in which he lived, and the satisfaction he derived from having led an useful as well as a happy life.

On Sunday he was much weaker, and only sat up in an armed chair while his bed was made. He desired me to read to him the eleventh chapter of John. I was going on to read to the end of the chapter, but he stopped me at the 45th verse. He dwelt for some time on the advantage he had derived from reading the scriptures daily, and advised me to do the same; saying, that it would prove to me, as it had done to him, a source of the purest pleasure. He desired me to reach him a pamphlet which was at his bed's head, Simpson on the Duration of future Punishment. " It will be a source of satisfaction to you to read that pampnlet," said he, giving it to me, " It contains my sentiments, and a belief in them will be a support to you in the most trying circumstances, as it has been to me. We shall all meet finally : we only require different degrees of discipline, suited to our different tempers, to prepare us for final happiness." Upon Mr. —— coming into his room, he said, " You see, Sir, I am still living."

Mr.

Mr. —— observed, he would always live. " Yes," said he, " I believe I shall ; and we shall all meet again in another and a better world." He said this with great animation, laying hold on Mr. ——'s hand in both his.

Before prayers he desired me to reach him three publications, about which he would give me some directions next morning. His weakness would not permit him to do it at that time.

At prayers he had all the children brought to his bed-side as before. After prayers they wished him a good night, and were leaving the room. He desired them to stay, spoke to them each separately. He exhorted them all to continue to love each other. " And you, little thing," speaking to Eliza, "remember the hymn you learned ; ' Birds in their little nests agree,' &c. I am going to sleep as well as you : for death is only a good long sound sleep in the grave, and we shall meet again." He congratulated us on the dispositions of our children ; said it was a satisfaction to see them likely to turn out well ; and continued for some time to express his confidence in a happy immorality, and in a future state, which would afford us an ample field for the exertion of our faculties.

On

On Monday morning, the 6th of February, after having lain perfectly still till four o'clock in the morning, he called to me, but in a fainter tone than usual, to give him some wine and tincture of bark. I asked him how he felt. He answered, he had no pain, but appeared fainting away gradually. About an hour after, he asked me for some chicken broth, of which he took a tea-cup full. His pulse was quick, weak, and fluttering, his breathing, though easy, short. About eight o'clock, he asked me to give him some egg and wine. After this he lay quite still till ten o'clock, when he desired me and Mr. Cooper to bring him the pamphlets we had looked out the evening before. He then dictated as clearly and distinctly as he had ever done in his life the additions and alterations he wished to have made in each. Mr. Cooper took down the substance of what he said, which, when he had done, I read to him. He said Mr. Cooper had put it in his own language; he wished it to be put in his. I then took a pen and ink to his bed-side. He then repeated over again, nearly word for word, what he had before said; and when I had done, I read it over to him. "That is right; I have now done."

About

About half an hour after he desired, in a faint voice, that we would move him from the bed on which he lay to a cot, that he might lie with his lower limbs horizontal, and his head upright. He died in about ten minutes after we had moved him, but breathed his last so easy, that neither myself or my wife, who were both sitting close to him, perceived it at the time. He had put his hand to his face, which prevented our observing it."

The above account, which conveys but a very inadequate idea of the composure and chearfulnees of his last moments deserves the attention of unbelievers in general, particularly of Philosophical Unbelievers. They have known him to be zealous and active in the pursuit of Philosophical truths and to be ever ready to acknowledge any mistakes he may have fallen into. By the perusal of these Memoirs they have found that he gradually, and after much thought and reflection abandoned all those opinions which disgrace what is usually called christianity in the eyes of rational men and whose inconsistency with reason and common sense has most probably been the cause of their infidelity and of their total inattention to the evidences of christianity. These opinions he abandoned

doned, because he could not find them supported
either in the Scriptures or in the genuine writings of
the early christians. They must be sensible that
the same desire for truth and the same fearless spirit
of enquiry and the same courage in the open avowal
of the most obnoxious tenets would have led him to
have discarded religion altogether had he seen reason
so to do, and there is little doubt but that he would
have been subject to less obloquy by so doing
than by exposing the various corruptions of christianity in the manner he did. They have seen
however that in proportion as he attended to the subject his faith in christianity increased and produced
that happy disposition of mind described in these
Memoirs. The subject is therefore well deserving
of their attention and they should be induced from so
fair an example, and the weight due to my father's
opinions, to make themselves fully acquainted with
the arguments in favour of christianity before they
reject it as an idle fable.

Many unbelievers have, no doubt, borne with
great patience severe calamities ; they have suffered
death with great fortitude when engaged in a good
cause, and many have courted death to serve their
friends

friends or their country. It must however be allowed that there is no great merit in meeting death with
fortitude when it cannot be avoided, and likewise that
the above cases cannot be absolutely calculated upon,
as there is no sufficient motive to account for their
conduct. But upon a truly practical christian there
is the greatest dependance to be placed for acting
well in all the situations in which he may be found,
his highest interest being connected with the performance of the greatest duties ; and even supposing
that many persons, who are not christians, from
favourable circumstances attendant upon their birth
and education, and from a naturally happy temperament of body and mind, may, and, it must be allowed
do acquire a habit of disinterested benevolence and
may in general be depended upon to act uniformly
well in life, still the christian has a decided advantage
over them in the hour of death, as to consider death
as necessary to his entering upon a new and enlarged
sphere of activity and enjoyment, is a privilege that
belongs to him alone.

APPEN-

APPENDIX, NO. 1.

Of the discoveries in factitious Airs before the time
of Dr. Priestley, and of those made by himself.

DR. PRIESTLEY has given a general though
brief account* of what had been done by his predecessors in this department of experimental Philosophy, and Sir John Pringle in his discourse before
the Royal Society on occasion of presenting Dr.
Priestley with the Copley Medal in 1772† has entered expressly, and more fully into the history of
pneumatic discoveries. The same subject was taken
up about three years after by Mr. Lavoisier still
more at large, in the introduction to his first Vol.
of Physical and Chemical Essays, of which a translation was published by Mr. Henry of Manchester in
1776. It is unnecsssary to detail here what they
have written on the history of these discoveries. It
may

---

* In the beginning of his first vol. of experiments : it is an abridgment of Sir J. Pringle's discourse.

† Discourses p. 4.

may be observed that no mention is made by any of these gentlemen of an experiment of Mr. John Maud, in July 1736*, who procured (and confined) inflammable air from a solution of Iron in the vitriolic acid. Inflammable air had been procured from the White Haven coal mines, and exhibited to the Royal Society by Mr. James Lowther, but I do not recollect any notice of its having been collected from a solution of metals in acids, and its character ascertained before Mr. Maud's experiment; for Hales, though he procured both inflammable and nitrous air, did not examine their properties. But it is much more extraordinary that neither Sir John Pringle who was a Physician, or Mr. Lavoisier who was so much occupied under government, respecting the Theory of the formation, and the practice of manufacturing Saltpetre from Nitre beds, should not have known, or have noticed the five treatises of Mayow on chemical, phisiological and pathological subjects, published a century preceding. Mayow

* Martyn's abridgment of the Philosophical transactions v. 9. p.396. I think Maud's experiment in 1736 likely to have suggested those of Mr. Cavendish in 1766.

yow is quoted by Hales,* by Lemery,† and by
Brown-

---

* Vegetable Statics v. 2. p. 234.

† Mem. de l' Acad. Royale 1717 p. 48.   On ne dit pourtant point
trop sous quelle forme ce nitre se contient dans l'air, et Mayou, Auteur
Anglois et grand defenseur du Nitre-Aèrien voulant èclaircir cette
difficultè, suppose l'air impregnè par tout d'une espece de nitre me-
taphysique, qui ne merite pas trop d'être refutè, quoi-qu'il l'àit ce-
pendant ètè suffisamment par Barchusen et par Schelhamer.   Le
fondement de l'opinion du Nitre aèrien, c'est comme le rapporte
Mayou lui même, qu'apres avoir enlevè à une terre tout le Nitre
qu'elle contenoit, si on l'expose ensuite à l'air pendant un certain
tems elle en reprend de nouveau :   il est vrai que si l'observation
ètoit parfaitement telle qu'elle vient d'être rapportèe, on auroit une
plus grande raison qu'on n'en a, de supposer dans l'air une très-grande
quantite de nitre, et de mettre sur le compte de ce nitre aerien un
grand nombre d'effets auquels il n'a certainement aucune part.

The experiment of Lemery mentioned in Dr. Watson's Essay on
Nitre, is in p. 54 of the Mem. de l'acad. royale for 1717 not for
1731.

It sometimes happens to men whose genius far transcends the
level of their day, to be from that very circumstance neither understood nor believed by their contemporaries.   Until the discoveries of
modern chemistry, who would have given Sir Isaac Newton credit for
his conjecture that the Diamond was an inflammable substance ? The
fact which Lemery sneers at, the reproduction of nitre in the earth, is
established beyond contradiction by the authors quoted by Dr. Watson

P

Brownrigg,* but though they appear to have read his work, it is evident that they knew not how to appreciate, or to profit by it. Haller† also refers to

---

son (Chem. Ess. v. 1 p. 318—321) and in Bowle's account of the nitre earths in Spain, and in Andreossi's memoir on the Saltpetre of Egypt. Though it is far from improbable that after lixiviation these earths may again become gradually impregnated with putrefying animal or vegetable matter to serve for the future crops of nitre.

* Philosophical transactions v 55 p. 232.

† Dr. Priestley in his preliminary account of the discoveries and theories on respiration (Exp. on air v. 3 p. 356. abridged edit.) quotes Haller's great work on Physiology. Haller quotes Mayow in three or four places ; but it is no wonder the quotations did not strike Dr. Priestley with any curiosity to examine Mayow's book, for Haller certainly did not understand his theory. For instance Lib. 8. § 13. Nitrum aereum. Si ad verum sensum nitri aerei hypothesis revocata fuisset parum utique ab eâ differt quam novissimê proposuimus. Nitrum quidem ipsum incautiosius olim Physiologi in aere obvolitare scripserunt, et ex pluviâ et nive colligi ; idemque passim ex rupibus efflorescere (Sprat ex Henshaw p. 264 major cal. hum.) exque plantis et stercoribus educi (Fludd Niewentydt, 563-4. Mayow de nitro aereo. Lower de Corde c. 3. Thurston 52. 53. Besse Analyse tom 1 et en lettre en reponse à M. Helvet. 114.) id nitrum aiunt in pulmonibus ad sanguinem venire, et ab eo ruborem illum elegantem, et fermentationem (Mayow, Thurston penult. ess. T. 3 p. 265 et calorem sanguinis cedere aut vicissim sanguinem condensari.

Certainly

to him, and he is respectfully quoted by Blumen-
bach* : but his book nevertheless long remained
in comparative obscurity.   From their time Mayow
has been neglected until his writings were noticed
by Dr. Forster, in 1780,† and again announced

as

Certainly the id nitrum, is not Mayow's.   M. Rosel seems first to
have ascertained the existence of nitre in plants.   A late experiment
of Dr. Priestley's, of which he gave an account in a letter to Dr.
Wistar, seems to make it probable that there may be nitre in snow.

* Blumenbach's Physiology, Caldweil's translation, Philadelphia, 1795.
§ 162.   Speaking of the theories of animal heat, " But all these hypotheses are embarrassed with innumerable difficulties ; whereas on
the other hand the utmost simplicity, and an entire correspondence
with the phenomena of nature combine in recommending and com-
firming that doctrine in which the lungs are considered as the focus
or fire place where animal heat is generated, and the deplogisticated
part of the air which we breathe as the fuel that supports the vital
flame.   That justly celebrated character Jo. Mayow sketched
out formerly the leading traces and the first great outlines of this
doctrine which in our times has been greatly improved, extended and
farther elucidated by the labours of the illustrious Crawford."

Dr. Darwin however is certainly right in supposing that heat is
evolved in many other processes of the animal economy, beside inspiration.

† See the translation of Scheele by Dr. John Reinhold Forster
1780 p. XIII.

In p.

as almost a discovery in the chemical world, by
Dr. Beddoes in the year 1790. His doctrines touch
so nearly on the subsequent discoveries of Priestley,
Scheele, Lavoisier, Crawford, Goodwin, &c. that it
seems absolutely necessary to discuss his pretensions, before those of his successors can be accurately admitted. As I am acquainted with Dr. Beddoes's pamphlet on Mayow, from the analytical review of it only, (V. vi.) and have no opportunity
here of consulting it, I shall take up Mayow's book,
and give an account of his tenets, from the work
itself.

Two of Mayow's Essays, viz. de Respiratione
and de Rachitide, appear to have been published at
Leyden, in 1671, the author who died at the age of
34, being then 26 years old. The propositions which
I have thought it necessary to extract from Mayow's
work, (ed. of 1674, Oxford,) and which I shall insert,
will give a concise, but faithful view of his discoveries

In p. 437 of v. 5 of the analytical review of Hopson's Chemistry,
before Dr. Beddoes's account of Mayow in 1790 the latter is stated
as the author of discoveries that might have given rise to the present system of pneumatic Chemistry.

coveries and conjectures in pneumatic Chemistry.* The abridgements of Beddoes and Fourcroy, I have no opportunity to consult, and as Mayow's book is far from being common, I have deemed it by no means an unnecessary labour to give the reader an opportunity of judging for himself, what is the precise extent of the claim, which the patrons of Mayow's reputation may fairly set up. It is also, of the more importance in a history of this subject, to notice the pretensions of this writer, as it appears that Boyle's experiments on artificial air, in his physico-mechanical experiments were not made until the year 1676 et seq. Though the first edition of that treatise repeatedly quoted by Mayow was in 1661. Mayow's experiments therefore ought to have been, and probably were known to Boyle at the publication of his last edition.†

The

---

* I believe Dr. Beddoes gives no more than the heads of each chapter and, a brief analysis of the contents. Dr. Beddoes in his remarks on Fourcroy's account of Mayow, Ann. de Chimie. No. 85, Nich. Jour. v. 3 quarto p. 108 states Mayow at the time of his death to have been only 27 and 28: but he was born in 1645 and died in 1769. Biog. Dict. 8vo. ed. of 1798.

† I do not find that Boyle quotes Mayow, though their labours in

the

The following is an analysis of Mayow's essays,
so far as relates to his chemical Philosophy.

CHAP. 1st. Of Nitre. The air is impregnated
with a vital, igneous, and highly fermentative spirit
of a nitro-saline nature. p. 1.

Nitre is a salt consisting of an acid and an alkaline part, as appears by the Analysis, and by the generation of nitre ; for if this salt be deflagrated with
sulphur, the acid spirit will fly off, and may be collected by means of a tubulated retort and a receiver :
and so if it be deflagrated with tartar, the residuum
will be equal in weight to the tartar employed,
though much of that, is of a fœtid oily nature.
This appears also from the composition of nitre, by
the addition of spirit of nitre to an alcali, p. 2–4.
The fixed part of nitre is obtained from the earth ;
                                                    pure

_____

the same field were contemporary.   But Boyle in his hidden qualities
of the air published in 1674 has an observation that looks as if derived
from Mayow. " And this undestroyed springiness of the air, with
the necessity of fresh air to the life of hot animals, suggests a great
suspicion of some vital substance if I may so call it, diffused through
the air, whether it be a volatile nitre or rather some anonymous substance, sidereal or subterraneal, though not improperly of kin to that
which seems so necessary to the maintenance of other flames."

pure earth being probably a compound of salt and
sulphur. p. 8.

CHAP. 2d. *On the aereal and fiery spirit of nitre.*
The air seems to contain an acid, as appears from
the regeneration of vitriolic acid after the calcination
of Vitriol, and from the rusting of steel filings in a
moist air; p. 10. A component part of the acid of
nitre, is derived from the air, which evidently contains something necessary to the support of flame.
But this aereal pabulum of flame, is not air itself,
for air remains when the confined taper is extinguished: nor is it as vulgarly supposed, the selt
called nitre, p. 12. But that these fire-air particlas
exist also in nitre is evident, since this salt will support the combustion of sulphur in vacuo. Fill
a tube with gunpowder slightly moistened, and it
will burn out in vacuo, or with its mouth inverted
over water. Hence the aereal part of nitre, is the
same with the fire-air particles of the atmosphere,
and is one component part of the acid spirit of nitre :
the other being (like the fixed part) obtained from
the earth, p. 17. 18. The fiery particles thus common to nitre and to the air, he denominates nitro-
aereal. It is these that give causticity to spirit of
P 4 nitre,

nitre, and occasion the red fumes observed in distilling it, p. 18.   They do not take fire of themselves in nitre, because they are inveloped with moisture; but when combined with salt of tartar, and thrown on the fire in a dry state they inflame, p. 20.

CHAP. 3*d*.   *Of the nature of the nitro-aereal and fiery spirit.* Fire he conceives to consist of these nitro-aereal particles set in violent motion by means of sulphureous bodies, in the cases of culinary fire : but by some other means, in the cases of the solar rays collected by a burning glass, and of the celestial fires.   The corrosive and caustic nature both of fire and nitrous acid, seems to argue that it proceeds in both from the nitro-aereal particles they contain, 22–24.   That fire is not of a sulphureous nature is evident, for nitre will not take fire in an ignited crucible; but oil thrown i , takes fire immediately.   So if a piece of metal be held over a candle, the fire particles pass through the metal, but the sulphureous smoke adheres to the under side.   p. 27.

That the heat occasioned by a burning glass, consists of these nitro-aereal particles is evident, for diaphoretic antimony may be made, either first by

calcina-

calcination with a lens, or secondly, by the repeated affusion of nitrous acid, or thirdly, by the deflagration of nitre on the antimony.   Diaphoretic antimony made by calcination, increases on weight,* by means of the nitro-aereal particles fixed in it by the process.   p. 28. 29.

CHAP. 4*th. On the origin of acid liquors, and the earthy part of Spirits of nitre.* From p. 34, it appears that he knew nothing of the absorption and combination of his nitro-aereal particles in the vitriolic acid, during the combustion of sulphur, but explains the whole mechanically by the saline portion of the sulphur being broken down into minute pointed particles, by the violent attrition of the nitro-aereal particles, and so becoming fluid and sharpened.   He seems too, not to know that the colcothar of martial vitriol is no component part of sulphur, p. 37.   The same mechanical explana-

---

* It was first observed by John Rey in 1630 that metals calcined, gain weight by the absorption of air.   See an account of his book by M. Bayen Journ. de Rozier 1775 v. 1 p. 48.   There are also some experiments by Boyle that shew the accession of weight on the calcination of metals, but he does not seem aware of the theory.   Shaw's Boyle, Fire and Flame weighed v. 2 p. 394, &c.

planation he applies to the formation of the ligneous acids, and to the impregnation of the caput mortuum or colcothar of vitriol, with fresh acid by exposure of air. In the succeeding paragraph, p. 39, he supposes that marchasite (martial pyrites) imbibes the mtro-aereal particles from the atmosphere, and thus acid is formed. In like manner he explains the formation of acids produced by fermentation, by the collision between the nitro-aereal, and the sulphureo-saline particles of the mass. p. 41. So also he supposes nitrous acid to be produced by the deten. tion of his nitro-aereal particles by the terrene saline particles found in the earth, p. 43. Hence he concludes generally, p. 43, that acid salts are formed from a saline basis brought into fusion or fluidity by the nitro-aereal part of the air : and sums up his theory of nitre, by stating it to be a triple salt, composed of nitro-aereal particles, united to a terrene basis forming the acid, which then unites to the fixed basis, supplied also by the earth.

Chap. 5th. On Fermentation. He gives in this chapter his theory of fermentation, as arising from the conflict of his nitro-aereal principle which he thinks may be termed mercury, and the sulphure-

ous

ous principle : evidently meaning by the latter, the Phlogiston of Stahl : and he states broadly, p. 60. that pure sulphur can never admit of accension, but by means of the nitro-aereal particles obtained from the atmosphere. The rest of his reasoning in this chapter, does not seem deserving of further notice.

CHAP. 6*th*. *On the nitro-aereal spirit as the cause of rigidity and elasticity.* These he explains by the fixation and state of his nitro-aereal particles in bodies endowed with these properties. In p. 69 he endeavours to account why boiled water freezes sooner than that which has not been boiled; a fact which Dr. Black has made the subject of a paper in the 45th vol. of the Philosophical transactions. But his reasonings throughout this chapter are not calculated to add to his reputation, or to the mass of knowledge of the present day.

CHAP. 7*th*. *The elastic force of the Air depends on its nitro-aereal particles. In what way exhausted air is reimpregnated with them. Of the elements of Heat and Cold.* This chapter contains experiments to shew that the elasticity of

of the air is owing to the nitro-aereal particles
contained in it: which may be destroyed by the
burning of a candle or other combustible substances, and also by the breathing of animals.
When the atmospheric air contained in a glass jar
inverted over water, will no longer support flame
or animal life, the water rises in the jar, owing to
the diminished elasticity of the air, not being able to
counter ct the pressure of the surrounding atmosphere on the water p. 100.   He finds p. 101 that
the diminution by burning a taper in a given quantity of the air, is about one thirtieth of the whole,
and by the breathing of mice and other animals
about one fourteenth.   Thence he concludes p. 106
that by means of respiration the elastic part of the
air enters into the blood, and that the sole use of
the lungs is not as some suppose, to break down the
blood in its passage into very minute particles.   That
combustion and respiration have similar effects on
atmospherical air, he concludes, p. 108, from the
fact, that a candle and a small animal inclosed together in a glass jar over water, the one will not burn,
nor the other remain alive above half the time that
they would if alone.   Mayow however, did not consider

sider his nitro-igneous and elastic particles to be either pure air, or even a component part of the common air, as air, notwithstanding the ambiguity of
the passages in p. 114 and 118 ; but as particles of
a different nature, attached to and fixed in the atmospheric particles ; and detached *(excussas)* by the
means above mentioned, p. 118 and 121.   His explanation of elasticity generally in this chap. and of
the difficulty arising from the obvious resistance to
the Atmosphere, and the expansibility of the air in
which a taper has been extinguished, or an animal
died, seem too obscure and unintelligible to merit
transcribing.   It is evident however upon the whole
from p. 123 compared with p. 100 and 135 that he
conceived the diminution of such air to arise from
diminished elasticity, but he supposes it to be den-
ser than common air 123.   In a subsequent part
of this chapter p. 128 et seq. he states his theory of
the manner in which deteriorated air recovers its
loss, viz. that the nitro-aereal particles being lighter
than the atmospherical, float abundantly in the higher
regions ; and that the part of the atmosphere deprived of them below, being forced upward by the
pressure of the atmosphere above, obtains a renewal
of

of these particles by mixture with the strata where they abound.

The element of fire, he supposes to reside in the body of the Sun, which is no other than a mass of nitro-aereal particles driven in perpetual gyration with immense velocity. Cold, which he considers as some thing positive (p. 130) he thinks consists in these particles assuming a pointed form, and moving not in gyration but strait forward. Much of his reasoning indeed throughout the book, savours greatly of the mechanical and corpuscular philosophy prevalent in his day.

CHAP. 8th. *On the nitro-aereal spirit as inspired by animals.* Formerly he thought that in respiration the nitro-aereal particles were rubbed or shaken off (*atterere, excutere* 146) from the common air by the action of the lungs, at present he thinks the air itself enters the mass of the blood, is there deprived of these particles, and of part of its elasticity. To prove this he produces an experiment of the diminution of air by the vapours from iron dissolved in nitrous acid: but the beautiful deductions of Dr. Priestley from a similar experiment, never occurred to him; on the contrary he expressly states that it

is

is an Aura, but not Air p. 145 and though afterward in chap. 9 p. 163, 164 he inclines to doubt,
yet again in p. 168 he denies it that character.

In p. 146 he proceeds to state the uses of these
nitro-aereal particles, which (147) he considers as
the principle of life and motion both in animals and
vegetables.   By the mutual action of the nitro-aereal,
with the sulphureo-saline particles contained in the
blood, a fermentation is excited necessary to animal
life, and to the warm fluid circulation of the blood
(*ad sanguinis æstum.*)   To these particles imbibed
from the air, he attributes the difference in colour between the venous and arterial blood ;   and he shews
this, from the numerous air bubbles arising in an
exhausted receiver from warm arterial blood : but
his experiment to illustrate the difference, from the
colour produced by the nitrous acid with vol. alk.
seems very little to the purpose p. 150.

To the fermentation arising from this mixture of
nitro-aereal particles with the blood, he ascribes animal heat, and accounts satisfactorily for the increased
heat of the body during strong exercise, from the
more frequent inspirations occasioned by the exertion (p. 152, 306 :) but his replies to the objections
                                                        of

of Dr. Willis, drawn from the phenomena of fer-
menting mixtures, are very inconclusive.

    Chap. 9th.   *Whether air can be generated anew.*
He repeats the experiment of dissolving iron in dilute
nitrous acid, and finds that though some of the vapour be absorbed, a portion still remains uncondem-
sible even by severe cold.   On substituting dilute
vitr. for nitr. acid he finds an aura which is hardly
absorbed or condensed at all.   Hence he doubts
whether these auræ be not entitled to the appellation
of air, especially as by subsequent experiment he
shews that they are equally expansible with common
air.   In making this last experiment he exhibits the
method of transferring air from one vessel to another
(Tab. 5. Fig. 5.) much in the manner afterwards
described by Mr. Cavendish in 1766.*   From the
inability of these auræ to support animal life (Tab. 5.
Fig. 6.) he concludes finally that they are not air,
though not very dissimilar p. 171. The succeeding five chapters do not seem to contain any facts or
conjectures that can add to Mayow's reputation.

                                         His

---

* Boyle had invented an apparatus for transferring air from one
receiver of an air-pump to another, but not under water.

His Hypotheses are completely superceded by the more accurate knowledge of the present day. In his tract on quick lime p. 225 he seems to have forestalled the acidum pingue of Dr. Meyer publish- ed exactly a century afterward. It may be noted that in his treatise on the Bath waters p. 259, he de- scribes fishes as collecting vital air from the water, and respiring like land animals. (Aereum aliquod vitale ab aquà, veluti aliàs ab aurà secretum et in cruoris massam trajiciatur.) The air bladder he considers rather as a reservoir of air to be inspired, than a re- ceptacle for excreted air ; though the latter opinion is made probable by Dr. Priestley.*

The first part of his *Treatises on Respiration* is is chiefly anatomical. In p. 300 et seq. he states more fully his opinion, that vital air, is of a nitro- saline nature : that it is the principle of life, both in Animals and Vegetables : that combined with the sulphureo-saline particles in the blood, it is the sti- mulus to the muscular fibre, and of course to the

heart

---

* See Nich. Journ. v. 3 p. 119 on the probability of fishes separating oxygen from the water they inhabit.

Q

heart as a muscle, p. 305 ; but that the fermentation occasioned by the introduction of these particles into the blood, is not confined to the left ventricle of the heart, but commences, in the passage of the blood through the lungs, and continues in the Arteries. This evidently approaches the theory, advanced by Dr. Goodwyn in his tract on the Connection of life with respiration about sixteen years ago, viz. that the pure air combined with the blood is the stimulus to the left ventricle of the heart, and produces the alternate contraction, and dilation on which the circulation depends. Dr. Lower, in his treatise de motu sanguinis, and Fracassati, and Dr. Frederick Slare attributed the change of the colour of venous blood into a florid red, to the combination of the air with it. Lower I believe preceded Mayow, who quotes him, p. 148; the date of Fracassati's and Dr. Slare's observations I have not been able to ascertain, but they must have been near the time of Mayow. Lowth. Ab. v. iii. p. 237.

In his third treatise on respiration, he explains the Animal œconomy of the fœtus in utero, by suggesting that the fœtus is supplied by the placenta, not

with

with venous, but with arterial blood brought by
the umbilical Arteries; so that the required stimulus of the nitro-aereal particles being thus conveyed,
supercedes the necessity of the lungs for the purpose.
This he ingeniously illustrates by the known experiment, that a dog into whom arterial blood is infus-
ed, though respiring with great difficulty before,
hardly respires at all. A similar theory he applies
to the life of the chick in ovo. This treatise seems
to have suggested Dr. Beddoes's illustration of his
theory of consumption from the state of pregnancy.

In a subsequent Essay on animal spirits, he conceives them to be, if not the same with the nitro-aereal part of the atmosphere, yet to consist of this, so
far as they are necessary to the production of muscular motion, which he attributes entirely as before to
nitro-aereal particles, p. 24 and 40, of chap. 4, on the
animal spirits.

I do not observe any thing else in Mayow's book
worth noting on the present occasion; or sufficiently connected with pneumatic Chemistry.

From the analysis thus given of* what Mayow
has

---

* At the time this was written neither Dr. Bostock's treatise on
respira-

has advanced, it appears, that he clearly comprehended the atmosphere to consist of a mixture of two parts, the one the efficient cause of life and of combustion, the other not of itself necessary to either.

That the vital part of the air, was also a constituent part of nitre, the effects of both being in essential particulars the same.*

That the vital part of the atmosphere entering the blood through the vessels in the lungs, is conveyed to the left ventricle of the heart, and becomes the stimulus to the contractions of that muscle, and is equally essential to the whole system of muscular contraction.

That

-----

respiration or the books therein quoted p. 200 had arrived here. Nor have I had an opportunity of consulting the references there made to Prof. Robinson, Dr. Thompson, Dr. Yeates, or Fourcroy's account of Mayow.

* Mr. Ray wrote " A dissertation (in 1696) about respiration," in which he supposes the air to pass from the bronchia and lungs into the substance of the blood, and there (pabuli instar) it foments and maintains the vital flame which he supposes to be in the sulphureous parts of the blood, as the air foments the common flame of a candle, and that the nitre has nothing to do with it. See Durham's collection of Ray's letters.

That the vital part of the atmosphere thus combined with the blood becomes also the source of animal heat.

That this vital part is equally necessary to the fœtus in utero as to the adult, and that the use of the lungs in the former case is superceded by the functions of the umbilical artery and placenta; by means of which, blood already impregnated with the vital air, is conveyed to the fœtus.

That the respiration of fishes, is dependant on the particles of air mixed with watery element they inhabited.

That heat, flame, and combustion, depend on two universal principles, and the gentleness or violence of their mutual conflict: the one being a principle of inflammability universally diffused in combustible bodies, and the other the vital or igneous part of the atmosphere.

These propositions evidently touch upon the most brilliant of the pneumatic discoveries of the authors already quoted; and not a little extraordinary it is, that they should have remained so long unknown, unnoticed, and not understood.

The sulphur of Mayow is decidedly the Phlogis-

Q 3 ton

ton of Stahl; the fire air of the former is the fire air of Scheele, the dephlogisticated air of Priestley, and the Oxygen of Lavoisier.

The combination of oxygen with the blood by means of respiration, first discovered as was thought by Lavoisier, is clearly stated by Mayow; who has also forestalled the elaborate theories of Crawford on animal heat, of Goodwyn, on muscular stimulus and of Beddoes on the succedaneum for respiration in the fœtus.

Boyle, though he must certainly have known of Mayow, neither quotes him, nor uses, or improves on his experiments; though as I have already remarked, he seems to have had notions of the atmosphere much like those adopted by Mayow. Whether this neglect arose from the pride of birth, or the pride of knowledge, or the pride of age, (for Boyle was almost twice the age of Mayow) or from jealousy of Mayow's abilities, cannot now be ascertained. From that time until Hales published his statics in 1726, pneumatic experiments were neglected, and the mathematical philosophy which Newton's discoveries rendered fashionable, absorbed for many years the attention of men of Science, particularly in England.

The

The way in which Lemery, Hales and Brownrigg speak of Mayow, evidently shews that his theories were not understood, nor his merits appreciated.

That Mayow was unknown to Black and Cavendish until of late years, is highly probable at least, if not absolutely certain. Neither these philosophers, nor Dr. Priestley, could have passed over Mayow's book, without being struck with his ideas, and publicly referring to them in their chemical works.

That Dr. Priestley was unacquainted with Mayow is certain, from the limited extent of his reading at the early period of his experiments (from 1770 to 1776 or 1777,) in books of chemistry and theoretic physiology : from Mayow, not being quoted by any of the writers whose works Dr. Priestley would be likely to consult except Hales and Brownrigg, and not by them in a manner to induce any farther curiosity : from their being unnoticed by Black, Cavendish, Sir. John Pringle, and Lavoisier, in particular : from the custom that Dr. Priestley had of acknowledging the sources of his ideas in all cases where they originated from the discoveries of others, as in his references to Hales, Brownrigg, Cavendish,

Q 4 &c ;

&c ; and from his making no mention of Mayow in his express account of the labours of his predecessors on the subject of animal respiration. That both he and Sir John Pringle before the Royal Society in 1772 and 1776 should expressly treat the *history* of discoveries in which Mayow bore so distinguished a part, and omit noticing him altogether, had they known of his works, is incredible. It is evident that he was then an obscure writer, and not in repute, or he would have occurred to them ; or some of their philosophical friends would have suggested the propriety of referring to his publications.

Neither is it likely that Scheele would have been acquainted with Mayow's writings, though it is singular that he escaped the notice of Lavoisier who I believe was employed under government in the collection of essays on the theory and manufacture of saltpetre and in the superintendance of the saltpetre works, especially as Mayow was mentioned though disrespectfully by Lemery, in his paper on nitre before referred to. But there certainly is no evidence that Lavoisier obtained his ideas of oxygen and its combination with the blood from Mayow, or his theory of metallic calcination from Jean Rey, though

his

his obligations to Dr. Priestley have not been always acknowledged with the candour and liberality that men of science would expect from Lavoisier.

Mayow had more than ordinary discernment in comparing known facts, and drawing conclusions from them, but he does not appear to have had the talent of imagining decisive experiments, of varying them, of observing and noting all the natural phenomena attendant upon them, or sufficient industry in pursuing them. It is one thing to make a plausible conjecture, and another to verify it. Those alone are entitled to the honour of discoveries who not merely start the theory, but take the pains of pursuing it by experiments and resting it on the basis of well conceived and accurately ascertained facts, sufficiently numerous and varied to obviate the most prominent objections. Mayow has reasoned with great acuteness and conjectured with singular felicity, but he added little to the mass of philosophical KNOWLEDGE in his day. He composed and decomposed nitre and ascertained the existence of vital air in this substance as well as in the atmosphere, but he did not collect, exhibit, and examine it. He knew how to make artificial air from nitrous acid and iron, but all the

the extraordinary properties of this gas, remained
unobserved by him as well as by others until collected and imprisoned by Dr. Priestley, and exposed to
the question under his scrutinizing eye.   Indeed as
an experimentalist Dr. Priestley stands unrivalled.
The multiplicity of his experiments, their ingenuity,
their bearings upon the point in question, their general importance, and their fidelity, were never
equalled upon the whole, before or since.   Nor is it
any detraction from their merit with those who are
accustomed to experiment, that they hold out no
pretensions to that suspicious accuracy, which has
too often depended more upon arithmetical calcula-
tions than upon actual weight and measure.   The
many kinds of aeriform fluids discovered by Dr.
Priestley, the many methods of procuring them, the
skilfull investigation of their properties, the foundation he laid for the labours of others, the simplicity,
the novelty, the neatness, and the cheapness of his
apparatus, and his unequalled industry, have deservedly placed him at the head of pneumatic Chemistry.   Nor should it be forgotten that while he
thus outstripped his predecessors and contemporaries
in the field of experiment, it formed not as with
                                                      them

them the business of his life, but (among other branches of literature and philosophy successfully cultivated) the occupation of his leisure hours, the relaxation from what he deemed more important, more laborious, and more obligatory pursuits.

Before his time (excluding Mayow) Boyle had discovered that air might be generated, fatal to animal life. It was known that common air would only serve a certain time for the purposes of combustion and respiration. The mephitic exhalations from natural Grottoes had been remarked. Inflammable air both natural and artificial had been exhibited before the royal society. Hales had ascertained the presence of air in a great number of substances where it was not commonly suspected though he had not the skill to examine the properties of the air produced. Black had ascertained the presence of fixed air in limestone, and Brownrigg, Lane, and Venel had illustrated the theory of mineral waters. But it was the paper of Cavendish in 1766 on fixed and inflammable air produced from various substances by means of acids, fermentation and putrefaction, that first introduced a stile of experimenting in pneumatic chemistry, more neat, more precise, and scientific than had hitherto been known.

The

The attention of Dr. Priestley however to these subjects was not originally excited by the works of his predecessors, but by the *accident* of his proximity to a brew-house at Leeds, where of course fixed air (a subject that had attracted much attention about that time) would be produced in a large way. It was thus that one experiment led to another, until the fruits of his amusements were the discoveries on which his philosophical reputation is principally founded. It is no more than justice to his character to mention in this place, that of all men living he was the freest from literary deception and the vanity of authorship. He never claims the merit of profound investigation or great foresight, for discoveries that might easily have been so stated as if they had been the pure result of those qualifications, but which were in reality the offspring of accident and circumstance. He excites others to patient labour in the field of experiment, from observing that success does not depend so much on great abilities or extensive knowledge, as on patient attention, and perseverance; and that much of his own reputation was owing to the discovery of facts that arose in the course of his pursuits, the result of no previous theory, unlooked

for

for and unexpected.   In v. 3 p. 282 of his experiments on air he says " Few persons I believe have " met with so much unexpected good success as " myself in the course of my philosophical pursuits. " My narrative will shew that the first hints at least " of almost every thing that I have discovered of " much importance have occurred to me in this " manner.   In looking for one thing I have general-" found another, and sometimes a thing of much "·more value than that which I was in quest of. " But none of these unexpected discoveries appear " to me to have been so extraordinary as that I am " about to relate (viz. the spontaneous emission of " dephlogisticated air from water containing a green " vegetating matter) and it may serve to admonish " all persons who are engaged in similar pursuits, " not to overlook any circumstance relating to an " experiment, but to keep their eyes open to every " new appearance and to give due attention to it " however inconsiderable it may seem."*   To this candour of disposition, and the readiness with

which

---

* See also the 1st, vol. of his early edition of experiments on air p, 39.

which he acknowledged his mistakes and his oversights, even those who opposed his opinions bear honoura- ble testimony. " The celebrated Priestley himself " (says M. Berthollet in his reply to Kirwan on " Phlogiston p. 124 of the Eng. translation) often " sets us the example, by rectifying the results of " some of his numerous experiments."

Numerous indeed those experiments were as well as important: far too numerous to be parti- cularised here; though it may not be improper to call to the recollection of the reader some of the more interesting facts which we owe to Dr. Priest- ley, and the times of their discovery and commu- nication.

The first of his *publications* on pneumatic che- mistry was in 1772, announcing the method of im- pregnating water with fixed air, and on the prepara- tion and medicinal uses of artificial mineral waters; a discovery that domesticated much of the know- ledge that had heretofore been disclosed only in the works of learned societies; and that beautifully exemplified how much of the health and the pleasure of common life, might depend on the ingenious re- searches of men of science. Though this was the

first

first publication of Dr. Priestley on the chemistry of the airs, he had certainly commenced his experiments in this branch of Science, soon after his arrival at Leeds, and as early at least, as 1768. In the year 1771 he had already procured good air from saltpetre ; he had ascertained the use of agitation, and of vegitation as the means employed by nature in purifying the atmosphere destined to the support of animal life, and that air vitiated by animal respiration was a pabulum to vegetable life; he had procured factitious air in a much greater variety of ways than had been known before, and he had been in the habit of substituting quicksilver in lieu of water, for the purpose of many of his experiments. In his paper before the Royal Society, in the spring of 1772, which deservedly obtained him the honour of the Copley Medal, he gives an account of these discoveries. In the same paper he announces the discovery of that singular fluid nitrous air,* and its beautiful

---

* Honestly referring to Dr. Hales and Mr. Cavendish for any idea that might have remotely led to this discovery (See Obs. on air 1st ed. v. 1 p. 108) the discovery however was completely his own.

Dr. Priestley seems always to have thought nitrous air as convenient

tiful application as a test of the purity or fitness for respiration of airs generally. In the same paper he shews the use of a burning lens in pneumatic experiments, he relates the discovery and properties of marine acid air; he adds much to the little of what had been heretofore known of the airs generated by putrefactive processes, and by vegetable fermentations, and he determines many facts relating to the diminution and deterioration of air, by the combustion of Charcoal, and the calcination of of metals.

Soon after this, in confirmation of Sir John Pringle's theory of intermittents and low fevers being generally owing to moist miasma when people are exposed to its influence, he ascertained by means of

his

---

ent a substance for eudiometrical experiments as any of the later substitutes, viz. the liquid sulphurets   the combustion of phosphorus. The foundation of Mr. Davy's substitute, muriat or sulphat of iron saturated with nitrous air, was as Mr. Davy acknowledges first discovered by Dr. Priesley himself. See Nich. Journ. for Jan. 1802 p. 41. The different states of the solutions of iron in vitriolic acid have been ingeniously applied to the analysis of mixed gasses by Humboldt and Vauquelin.

his nitrous test that the air of marshes was inferior in purity to the common air of the atmosphere.*

He had obtained very good air from saltpetre in 1771, but his full discovery of dephlogisticated, air, seems not to have been made until June or July, 1774,† when he procured it from precipitate per se, and from red lead. This was publicly mentioned by him at the table of Mr. and Madame Lavoisier, at Paris, in October 1774, to whom the phenomena were until then unknown. The experiments on the production of dephlogisticated air, he made before the scientific chemists at Paris about the same time, at Mr. Trudaine's. This hitherto secret source of animal life and animal heat, of which Mayow had but a faint and conjectural glimpse, was certainly first exhibited by Dr. Priestley, and about the same time, (unknown to each other) by Mr. Scheele of Sweden. For the honour of science, it were much to be wished that the pretensions of Mr. Lavoisier were equally well founded. He has done sufficient

and

---

* Phil. trans. v. 54 p. 92.

† See Doctrine of Phlog. established p. 119,

R

and been praised sufficiently for what he has done, to satisfy a mind the most avaricious of fame ; he is deservedly placed in the first rank among the philosophers of his day, and he ought not to have thrown a shade over his well earned reputation, by claiming for himself the honour of those discoveries which he had learned from another.

From this brief account of the first stage of Dr. Priestley's chemical labours, it appears that during the short period of two years, he announced to the world more facts of real importance, and extensive application, and more enlarged and extensive views of the œconomy of nature, than all his predecessors in pneumatic Chemistry had made known before.

In 1776 his observations on respiration were read before the Royal Society ; in which he clearly discovered that the common air inspired, was diminished in quantity, and deteriorated in quality, by the action of the blood on it *through the blood vessels of the lungs ;* and that the florid red colour of arterial blood, was communicated by the contact of air through the containing vessels. His experiments on the change of colour in blood confined in a bladder, took away all doubt of the probability of this

mode

mode of action.   I cannot help thinking that the circumstance of Dr. Priestley's mind being so much
occupied with the prevailing theory of Phlogiston,
was the reason why he did not observe that the diminution of the air, and the florid colour of the arterial blood was owing to the absorption of the pure
part of the atmosphere, *rather* than to any thing
emitted from the blood itself.   This part of the theory of respiration Mr. Lavoisier has certainly established ; though it is by no means ascertained as yet
whether the vital part of the atmosphere inspired, is
wholly and alone absorbed, or whether in reality
something is not contributed in the lungs to the formation of the fixed air found after expiration.*

In 1778 Dr. Priestley pursued his experiments on
the property of vegetables growing in the light to
correct impure air, and the use of vegetation in this

part

---

* That azote is absorbed during respiration as Dr. Priestley supposed contrary to Mr. Lavoisier's opinion, is made extremely probable by the experiments of Mr. Davy, whose accuracy is well known.
Researches, p. 434. The formation of water in this process, is certainly no more than conjecture as yet.   Dr. Bostock has lately published a very useful and laborious history of discoveries relating to
respiration, both anatomical and pneumatical.

part of the œconomy of nature. A discovery which was announced to several men of science in England previous to the publication of the same ideas by Dr. Ingenhouz.* Indeed from its having been communicated to M. Magellan whose pleasure and whose occupation it was, to give information of new facts to his philosophical correspondents, and of this in particular to Dr. Ingenhouz then engaged in similar researches, there is hardly a doubt but the latter knew of the experiments then pending on the subject by Dr. Priestley.

It is painful to notice these aberrations from propriety in the conduct of men highly respectable in the philosophical world, arising from an over anxious avarice of literary fame, and an improper jealousy of the reputation of another. Not that it derogates from the character of a philosopher to wish for the

applause

---

* Doctrine of Phlogiston established, p. 107, et. seq. The theory of the amelioration of impure air by the absorption and excretion of vegetables growing in the light, has been doubted by Dr. Darwin in his Phytologia, and opposed by Count Rumford in a paper published in the transactions of the Royal Society, for 1787 : also by Dr. Woodhouse of Philadelphia, Nicholson's Journal, for July 1802, and by Mr. Robert Harrup, Nicholson's Journal, for July 1803.

applause of those who know how to appreciate his merit, or who are benefited by his exertions; such an anxiety is laudable when it does not lead to encroachments on the literary rights of others; nor is it at all desireable under the present circumstances of human nature, to expect from men of science an attention to their pursuits arising from motives of pure benevolence alone, and excluding all views, hopes, and expectations of the gratifying tribute of public approbation. I believe no man ever laboured with a more single eye to public utility than Dr. Priestley. But consideration in society, and the respectability attendant upon great talents, and great industry, successfully employed for the benefit of mankind, is a motive to useful exertion so universal, so honest, so laudable, and withal so powerful, that it is the common interest, as well as the duty of society, to bestow it liberally where it has been earned faithfully, and to concede it to those only, who have really deserved this honourable reward.

From this period Dr. Priestley seems to have attended to his pneumatic experiments as an occupation; devoting to them a regular portion of his time. To this attention, among a prodigious variety of

R 3

facts

facts tending to shew the various substances from
which the gasses may be procured; the methods
of producing them; their influence on each other,
and their probable composition, we owe the discovery of vitriolic acid air, of fluor acid air, of vegetable acid air, of alkaline air, and of dephlogisticated
nitrous air, or gazeous oxide of azote as it has been
called, the subject of so many curious experiments
by Mr. Davy. To these we may add the production of the various kinds of inflammable air by numerous processes that had escaped the observation
of Mr. Cavendish; in particular the formation of it
by the electric spark taken in oils, in spirits of wine
and in alkaline air; the method of procuring it by
passing steam through hot iron filings, and the phenomena of that hitherto undetermined substance
the finery cinder, and its alliance to steel. To
Dr. Priestley we owe the very fine experiment of
reviving metallic calces in inflammable air and its
absorption in toto, apparently at least, undecompos-
ed. He first ascertained the necessity of water to
the formation of the gasses, and the endless production of air from water itself.

Dr. Priestley's experiments on this subject, to
wit:

wit: the generation of air from water, opened a new field for reflection, and deserves more minute notice. No theory has yet been proposed adequate to the explanation of the facts. He had before remarked that water was necessary to the generation of every species of air, but the unceasing product of air from water had never been before observed.

In his first set of experiments he procured air, by converting the whole of a quantity of water into steam : then, to obviate the objection to the water having imbibed air from the atmosphere he put the water on mercury in long glass tubes immersed in mercury : in a third process he used no heat, but merely took off the pressure of the atmosphere. In all these cases a bubble of air was extricated from the water, which being separated by inclining the tube, another bubble was again produced on each repetition of the experiment. That this could not be air imbibed from the atmosphere appeared from this, that though the first portions were generally purer than atmospheric air, the next became less pure, and at length wholly phlogisticated.

It did not appear that the addition of acids, en abled the water to yield more air, nor did he suc-

R 4   ceed

ceed in attempting to convert the whole of a given quantity of water into air, although exposing the water confined over mercury to heat, and separating the air produced, it still continued to produce more air for twenty or thirty repetitions of the experiments. When a certain proportion of air was thus produced at any one time, no continuance of the experiment would encrease the quantity until it was separated. Hence he concludes that the longest continuance of of water in the state of vapour would not convert it into air. The water used was pure distilled water previously boiled to separate any adventitious air that might have been imbibed from the atmosphere. The precautions he used, and the replies to such objections as he foresaw the experiment would be liable to, are detailed in the papers he published on the subject, to wit, a separate pamphlet published in England in 1793, and a communication in the Am. Ph. trans. v. IV. p. 11—20.

In the last mentioned paper, he proceeds also to give an account of some experiments on the property of water to imbibe different kinds of air, and the conversion of sp. of wine, into inflammable air.

This paper inserted in the American transactions,

was

was read before that society in Feb. 1796.   In Ap.
1800 another paper was read before the same society
on the production of air by the freezing of water Am.
Ph. trans. v. V. p. 36.   In this paper he recapitulates
the general result of his former experiments on the
generation of air from water, namely " that after all
" air had been extracted from any quantity of water
" by heat or by taking off the pressure of the atmos-
" phere, whenever any portion of it was converted
" into vapour, a bubble of permanent air was formed,
" and this was always phlogisticated.   The process
" with the Torricellian vacuum (he says) I continued
" for some years and found the production of air
" equable to the last.   The necessary inference from
" this experiment is, that water is convertible into
" phlogisticated air, or that it contains more of this
" air intimately combined with it than can be ex-
" tricated from these processes in any reasonable
" time."

He proceeds to state his imperfect attempts to procure air from water by freezing, until he procured
cylindrical iron vessels seven or eight inches high and
near three inches wide at the bottom, the upper ori-
fice closed with a cork and cement, in the centre of
which

which was a glass tube about one fifteenth of an inch in diameter. In this apparatus the water in the iron vessel was frozen by means of snow and salt, the vessel being immersed in mercury, and the water contained over the mercury. The quantity of water was about three ounces. The experiment was repeated nine times without changing the water, and the last portion of air procured in this manner was as great as any of the preceding; so that there remained no reasonable doubt but that air might be produced from the same water in this manner ad libitum. Having obtained near two inches of air in the glass tube, Dr. Priestley put an end to the experiment, and examining the air found it wholly phlogisticated, not being affected by nitrous air, and having nothing inflammable in it.

The inference drawn by the Doctor from those experiments is, that water when reduced by *any means* into the state of vapour, is in part converted into phlogisticated air; and this is one of the methods provided by nature for keeping up the equilibrium of the atmosphere, as the influence of light on growing vegetables is the means of recruiting the other part; both of them being subject to absorption and diminu-

diminution in several natural processes. And he thinks that they strengthen also the opinion, that water is the basis of every kind of air, instead of being itself a compound of hydrogen and oxygen according to the new theory. At all events the experiments themselves must be considered as extremely curious, as well as new.

The water and the salt thus made use of gave rise to another experiment of the most important nature to the present theory of chemistry, if it should on future repetition be ultimately verified. This experiment related by Dr. Priestley in a letter to Dr. Wiston is in substance as follows. Having repeatedly used as above mentioned a freezing mixture of common salt and snow, the experiment being finished, he evaporated the snow water in an iron vessel and recovered the salt. The salt thus recovered contained some calx of iron. He put it by in a bottle and labelled it, according to his usual practice. In October 1803, he wanted to procure some marine acid, and took the salt thus procured by evaporating the snow water, for the purpose. On commencing the distillation, he was surprized to find the receiver full of the characteristic red fumes of the nitrous acid.

acid. The vitriolic acid used for the purpose was diluted with about an equal quantity of water. On finishing the process, he took some of the acid in the receiver, and dissolved copper in it, and thus procured good nitrous air. He was himself perfectly persuaded that no nitre had been used in the freezing mixture, nor had any by accident or design been mixed with the salt. He was not unacquainted with the common mode of clearing black oil of vitriol by the addition of nitre. So that no means of accounting for this curious fact remained, but the snow or the iron : he seemed to think that should this experiment be fully verified hereafter, it would confirm the vulgar hypothesis of snow containing nitre, and account for the fertilizing quality usually attributed to snow. He had no opportunity in that winter of repeating the experiment as he died in about three months after, and his previous illness had compelled him to forsake his laboratory.

Of the almost discarded theory of Phlogiston Dr. Priestley to his death remained the strenuous advocate, and almost the sole supporter; *ipse Agmen.* Beautiful and elegant as the simplicity of the new doctrine appears, many facts yet remain to be explained,

plained, to which the old system will apply, and the French theory is inadequate. These are collected with an ingenuity of arrangement, and a force of reasoning in the last pamphlet published by the Doctor on the subject,* which no man as yet unprejudiced can peruse, without hesitating on the truth of the fashionable theory of the day.

Certainly, it has not yet been sufficiently explained on the new theory, what becomes of the Oxygen from the decomposed water in the solution of metals in acids; nor why inflammable air is produced when one metal in solution is precipitated by another; nor why dephlogisticated air is hardly to be procured from finery cinder, if at all; nor why this substance so abounding in oxygen according to the new theory, will not oxygenate the muriatic acid; nor why it should answer all the purposes of water in the production of inflammable air from charcoal; nor why water in abundance should be produced when finery cinder is heated in inflammable air, and none when red precipitate is exposed to the same process; nor what becomes of the oxygen of the decomposed water

----

* The doctrine of phlogiston established 1803.

ter when steam is sent over red hot Zinc, and inflammable air is produced without any addition in weight
to the Zinc employed; nor why there should be a
copious production of inflammable air when hot
filings of Zinc are added to hot mercury in a hot
retort and exposed to a common furnace heat, which
I believe is an unreported experiment of Mr. Kirwan's; nor why sulphur and phosphorus are formed
by heating their acids in inflammable air without our
being able to detect the oxygen which on the new
theory ought to be separated, nor why water should
be produced by the combustion of inflammable air
with ,47 of oxygen, and nitrous acid when ,51 of
oxygen is employed, for this experiment can now no
more be doubted than explained; nor why on the
new doctrine the addition of phlogisticated air, should
make no alteration in the quantity of acid thus obtained; nor why red hot charcoal slowly supplied
with steam, should furnish inflammable air only and
not fixed or carbonic acid air; nor why nothing but
pure fixed air should be produced by heating the car-
bonated Barytes in the same way; nor why fixed
air should be formed under circumstances when it
cannot be pretended that Carbon is present, as when

gold,

gold, silver, platina, copper, lead, tin and bismuth
are heated by a lens in common air over lime water;
or why the grey and yellow calces of lead should furnish carbonic acid and azote, and no oxygen; nor
why the residuum of red lead when all its oxygen is
driven off by heat should be either massieot or glass
of lead according to the degree of heat, and not lead
in its metalline state; nor why plumbago with steam
should yield inflammable and not fixed air; nor why
minium and precipitate per se heated in inflammable
air should produce fixed air; nor why on the evapo-
ration of a diamond in oxygen, the fixed air produced
should far exceed the weight of the diamond employed, if some of the oxygen had not entered into the
composition of the carbonic acid so formed; nor
why there should be a constant residuum of phlogisticated air (or azote) after the firing of dephlogisticated and inflammable airs, if it be not formed in the
process; nor why phlogisticated air if a simple substance, should be so evidently formed in the various
processes enumerated by Dr. Priestley in the 13th
section of the pamphlet of which I have made the
foregoing abstract? whether the doctrine of phlogiston is still to be used as the key to the gate of chemical

mical theory, or whether it be properly thrown aside for the elegant substitute of the French chemists, can hardly be ascertained, until the preceding difficulties are cleared up on the new doctrine, for on the old theory they are sufficiently explicable. The summary of arguments in favour of Phlogiston, published by Dr. Priestley, in 1803, are evidently too important, and too difficult of reply, to be slighted by those who adopt the opposite opinions. *Non nostri est tantas componere lites.* Should the old theory ultimately fall, it may be fairly said of its respectable supporter, *si Pergama dextra defendi potuit, etiam hac defensa fuisset.*

This was almost the last of Dr. Priestley's chemical publications,* through all which, his characteristic talent as an author has been eminently preserved, that of not only adding greatly to the existing stock of knowledge, but exciting others to exertion and reflection in the same line of pursuit. Nor can

I help

---

* To the end of this Appendix will be subjoined a list of the scattered papers on Philosophical subjects which Dr. Priestley published in periodical collections, besides those which are inserted in the Philosophical transactions.

I help thinking that much of the labours of the French philosophers in this department of science would never have been undertaken, if they had not been called forth by the previous discoveries, not of Lemery, Margraaf, Bayen, Macquer, and Beaumè, but of Hales, Black, and Macbride; of Cavendish and Priestley and Scheele.* Would to God there were no other object of contest between the rival nations of Great Britain and France, but which should add most to the sum of human knowledge, and contribute most to the means of human happiness.

It is impossible to conclude the preceding account better than by the following extract of a letter to Mr. Lindsey from a man† well able to appreciate the labours of Dr. Priestley; and the late testimony in favour of his discernment by Dr. Bostock. " To " enumerate Dr. Priestley's discoveries, would in

" fact

---

* I do not mean to deny the tribute of praise to Marriotte and Venel, any more than to Brownrigg and Lane, and it is certain that Lavoisier was engaged in pneumatic experiments, previous to 1774.

† Richard Kirwan, Esqr.

S

" fact be to enter into a detail of most of those that
" have been made within the last 15 years. How
" many invisible fluids whose existence evaded the
" sagacity of foregoing ages has he made known to
" us? The very air we breathe, he has taught us to
" analyze, to examine, to improve : a substance so
" little known, that even the precise effect of respira-
" tion was an enigma until he explained it. He first
" made known to us the proper food of vegetables,
" and in what the difference between these and ani-
" mal substances consisted. To him Pharmacy is
" indebted for the method of making artificial mi-
" neral waters, as well as for a shorter method of
" preparing other medicines ; metallurgy for more
" powerful and cheap solvents ; and chemistry for
" such a variety of discoveries as it would be tedious
" to recite : discoveries which have new modelled
" that science, and drawn to it and to this country,
" the attention of all Europe. It is certain that
" since the year 1773, the eye and regards of all the
" learned bodies in Europe have been directed to
" this country by his means. In every philosophi-
" cal treatise, his name is to be found, and in almost
" every page. They all own that most of their dis-
                                        " coveries

" coveries are due either to the repetition of his dis-
" coveries, or to the hints scattered through his
" works.*

" This is not the only instance" (says Dr. Bostock,† speaking of Mr. Jurin's opinion that azote
was generated, instead of being absorbed, in the process of respiration as Dr. Priestley, and after him
Mr. Davy had supposed,) " in which, after the con-
" clusions of Dr. Priestley have been controverted
" by his contemporaries, a more accurate investiga-
" tion of the question, has ultimately decided in his
" favour. The complicated apparatus, and impo-
" sing air of minuteness which characterize the ope-
" rations of the French chemists, irresistibly engage
" the assent of the reader, and scarcely permit him
" to examine the stability of the foundation upon
" which the structure is erected. The simplicity
" of the processes employed by Dr. Priestley, the
" apparent ease with which his experiments were
" performed, and the unaffected conversational stile
" in

* Vindiciæ Priestlianæ, p. 68.

† Essay on respiration, p. 208.

" in which they are related have, on the contrary been
" mistaken for the effects of haste and inaccuracy.
" Something must also be ascribed to the theoreti-
" cal language which pervades, and obscures the
" chemical writings of this Philosopher, in conse-
" quence of his unfortunate attachment to the doc-
" trine of Phlogiston."

When the operose experiment of the French chemists on the formation of water, shall have been sufficiently repeated, and verified by other experiments
to the same point, less complex, less tedious, less expensive, and easy to be repeated; when the water
thus supposed to be formed is sufficiently distinguished from the water absolutely necessary to the
generation of all airs, and attendant upon them* both
in a state of mixture and combination; and when
the difficulties enumerated a page or two back, as attendant on the modern theory shall be explained on
the

---

* Mr. Kirwan found that common inflammable air from iron, and
vitriolic-acid, contained about 2-3 of its weight of water mixed with
it; which might be separated from the air by means of concentrated
vitriolic-acid in a watch glass over mercury, without diminishing the
quantity or altering the characteristic properties of the air thus
treated.

the new system, as well as on that of Stahl, then, and not until then, will it be time to lament Dr. Priestley's unfortunate attachment to the doctrine of Phlogiston.

———•+•———

Of Dr. Priestley's other Scientific Works.

THE other philosophical labours of Dr. Priestley consist of his history of electricity, his history of the discoveries relating to light and colour, and his popular introductions to perspective, electricity and natural philosophy.

It appears that after the publication of his history of electricity, he intended to have pursued the plan, by composing similar histories of every branch of science : a magnificent idea, and which none but a man conscious of uncommon powers could have contemplated. Few men indeed were so capable of such an undertaking as Dr. Priestley ; for independant of his habits of patient and regular industry in his literary pursuits, and the wide field of his attention to scientific objects, he had a facility of perusing, abstracting, and arranging the works of others, not commonly attendant even upon equal abilities in

S 3   other

other respects. This great undertaking of Dr. Priestley to embrace the various departments of philosophy, appears a labour sufficient for one life ; and had due encouragement been afforded, this projected series of histories would in all probability have been compleated, usefully to the world, and reputably to himself. But he proposed this undertaking laborious as it was, without designing that it should occupy the whole or the principal portion of his time, but his leisure hours only ; for at no period did he postpone his professional duties, or his theological studies, to any other object whatever. The life of Dr. Priestley is almost a perpetual illustration of a seeming paradox, respecting mental energy, that men of talents, uncommonly laborious, and who appear to get through more business than one person could be supposed equal to, have usually more leisure time at their disposal, than those who have little to do : so much does the habit encrease the power of exertion. Nor was any man less averse to the innocent pleasures of social enjoyment than Dr. Priestley, or better calculated as well as more inclined to contribute to the common stock of amusing, and instructive conversation. It cannot

not indeed be truly said of him, as Dr. Johnson*
once related of himself, that he had never refused an
invitation to dinner on account of business but once
in his life, yet no man more readily found leisure for
social intercourse. This arose from his habit of di-
viding his time into certain portions appropriated to
his respective pursuits, and determining to perform
a certain quantity of literary duty, within the assigned period.

The first edition of his history of Electricity, was
in 1767 : it went through another edition in 1769,

and

---

* On that day, (Dr. Johnson said) as it was an unusual deprivation,
he found himself disinclined, and unable to attend steadily to the
work that led him to refuse the invitation. He walked about his library occasionally looking over first one book and then another until
about four o'clock when weary of staying within he went to a tavern to
dine. Dr. Johnson had for a long time a dislike to Dr. Priestley
who bore two of the characters most in disrepute with Dr. Johnson, that
of a whig and a dissenter. Dr. Priestley's pursuits also consisting so
largely of heterodox theology, which Dr. Johnson abominated, and
experimental philosophy which he heartily despised, they had hardly
a common point of union. Toward the latter part of Johnson's life,
they met ; and upon the friendly terms that ought to obtain between
two men, who, each in their way, deserved so well of the republic of
letters.

and a third in 1775. It was published at a very happy time, when electricity was a favourite object of attention to many respectable men of science then living, and it contributed in a great degree to turn the public attention toward the study of these phenomena. Very much of what has been done since may be fairly attributed to the popularity given to this branch of experimental philosophy by Dr. Priestley. Nor did he confine himself to a mere narration of the labours of others; the second volume contains many new experiments of his own, and some of them form very curious and important additions to the stock of electrical knowledge.*

           The

---

* Dr. Priestley among his other experiments on electricity first ascertained the conducting power of charcoal and the calcination and vitrification even of the most perfect metals by the electric spark He seems first to have used large batteries, which M. Van Marum and his associates have carried to such extent.

The solutions of the metals, the gasses produced and the circumstances which accelerate and prevent these effects in Galvanic processes with the pile of Volta, as detailed by Dr. Priestley in his paper on this subject in Nich. Jonrn. for March 1802 p. 198 form very important additions to the mass of knowledge respecting the Galvanic-fluid. Nor are his discoveries in pneumatic electricity, of the conversion of oils, spirit of wine and the alkaline gass into inflammable air or hydrogen of less moment.

The discoveries of the last thirty years, particularly including those of Galvanic Electricity, are so numerous, and so dispersed in volumes difficult to be procured, that a continuation of this history is a desideratum in the scientific world ; at one time there was an expectation of seeing it from the pen of Mr. Nicholson, whose general knowledge, and industry, as well as his attention to this branch of philosophy in particular, render him peculiarly qualified for the task.   But the proposals he communicated to Dr. Priestley, on the subject, were not pursued to effect.*

These histories of detached branches of Science, would not only be highly useful, but they may be considered as in some measure necessary to the accurate pursuit, and advancement of science itself. They are not only useful for the purpose of shewing the discoveries that have been made, and the time of their publication, the ideas that appear to have suggested them, the persons to whom we are indebted for them, and their effect on the spirit of enquiry at

the

---

* Dr. Bostock, who seems to have many requisites to qualify him as the historian of particular branches of science, has published a good attempt toward the history of Galvanism in Nicholson's Journal.

the time, but they prevent a man of science from being led into mistakes, from doing what has been already done, from suggesting what has been already published, and from ignorantly claiming to himself the merit due to the labours of a predecessor. Books are now so multiplied, in languages so various, obtained with so much difficulty, and at an expence so far exceeding the usual means of scientific men, that those who like Dr. Priestley fully and faithfully execute a work of this description are real benefactors to mankind.*

The history of ELECTRICITY was composed by Dr. Priestley in one year. The three editions of the work in less than eight or nine years sufficiently shew that, in the opinion of men of science, it was well composed: otherwise the celerity of its composition, would no doubt derogate from, instead of adding to, the well earned reputation of the author; and rather tend to shew that he was too careless or too conceited to take the necessary pains and employ the necessary

time

---

* The transactions of the various academies and philosophical societies in Europe amount at least to 1000 volumes in quarto. The royal society of England in 1665 led the way to similar institutions.

time to make it fit for public inspection.   Every man
owes to the public, that if he professes to instruct
them, he should dedicate as much labour as the subject demands, or at least as much time as it is in his
power to devote to it.   I fully accede to the ingenious correction of the *nonum prematur in Annum*,
suggested by the witty Dr. Byrom of Manchester;
but something of the *Limæ Labor*, respect for the
tribunal of the public demands of every man who appears before them in the character of an author.   Dr.
Priestley has in more instances than one, been accused of unnecessary if not of culpable rapidity in his
literary compositions : but he never professed to be
a fine writer; he never sought after the beauties of
stile ; and his common language was sufficiently neat
and expressive, to communicate the facts and the arguments upon which it was employed.   It is also to
be remarked, that the facility of composition which
he acquired from long practice, made that labour
light to him, which would have been too much for a
less skilful and a less experienced composer.   In
many instances indeed of his rapid publications, he
had not to *seek* for arguments, but to express in his
unornamented and unaffected manner, the ideas that
forced

forced themselves upon him relating to a subject previously considered and upon which he had long made
up his mind.

The History of Discoveries respecting Light and
Colours published in 1772 was a more difficult task,
nor did it meet with equal encouragement.   Sir
Isaac Newton's important labours in this branch of
science, could not be fully comprehended without a
portion of mathematical knowledge not even then so
common as formerly, among the philosophers of the
day.   Mathematical studies seem to have in themselves very little to interest, compared with other literary pursuits; although by long attention and habit,
that interest may be excited and kept up.   It was
about this time that the popular phenomena of chemistry and electricity more decidedly took their stand
in the field of science, and irresistably seized hold on
the attention of the world : phenomena, highly amusing in themselves, strongly attractive from their novelty, of evident and immediate application, and
that promised an incalculable harvest of honourable
and useful discovery, to such as would become their
votaries.   Little had been done in this department of
philosophy, little previous knowledge was required

to

to comprehend all that was known, and those who were unable to read a page of Sir Isaac Newton with profit, could easily mix an acid and an alkali, or turn the wheel of an electrical apparatus.

By this time too, it had been discovered, that there were other powers in nature that must be called in to explain appearances, which the mechanical and corpuscular philosophy had endeavoured to elucidate in vain. Such were magnetism, electricity and chemistry. It began to be found out, that the science of calculation, was but an aukward handmaid to their sister branches of natural philosophy, while physiology, laughed outright at the clumsy addresses of her mathematical admirers, from Borelli to Keill.

The discoveries therefore relating to light and colours, at the time when Dr. Priestley proposed his history, being intimately associated with the study of the mathematics, and the profound investigations of Sir Isaac Newton, were out of the beat of the less laborious, but more fashionable philosophy of the day; and were not so generally interesting to the Sciolists and Amateurs. Hence the work in question, though treated in a very entertaining and popular manner, and by no means crouded with reference

ference to Diagrams or abstruse discussions, was not popular even among that class of readers, who might reasonably be calculated on, as the purchasers of such a performance. The subscribers indeed were sufficiently numerous, and respectable, but by far the majority were defaulters in respect of payment. It did not pay the bookseller: and of course still less did it recompence Dr. Priestley in a pecuniary point of view, especially as he had gone to considerable expence with a view to the completion of his extended plan. To him indeed, though pecuniary loss was a serious evil, pecuniary profit was a consideration of small importance: his motives to literary labour seem uniformly to have arranged themselves as follows, utility, reputation, profit.

The work in question is certainly too brief, considering the importance of the subject: many parts of it, the theory of Huygens, Euler, and Franklin for instance, seem to have merited more discussion. That all the phenomena of light depend on the Sun, as the reservoir, whence all the emanations of that fluid to the various parts of the system are supplied, the lighting of a candle is alone sufficient to refute. The facts discovered to us by modern Chemistry

will

will suggest a great many other doubts of the doctrines respecting light, which were regarded as well established when Dr. Priestley's book was written. But it was a faithful account of the knowledge of the day, and an unprejudiced tribute to the reputation of those philosophers who had from time to time extended the boundaries of science on the subjects treated of.

Not a little has been added to the mass of facts then published, by the subsequent experiments of Dr. Priestley himself, and his fellow labourers in the Chemistry of the Gasses : and notwithstanding the experiments of Sir Isaac Newton and his predecessors, the theory of light and colours is not yet rested upon facts sufficiently numerous, and decisive to satisfy the enquiries dictated by the present state of knowledge.

But with all these disadvantages, the work has nevertheless maintained its ground, for we have no where else so systematic, and compleat, though brief an account of what had been made known to the world on this important branch of scientific inquiry. It will always remain a valuable performance; and to the author an honourable one, from the knowledge

ledge and ability required in its compilation, from the fairness of the account it gives, and the entertaining statement of facts and suggestions interspersed through the book.

It is greatly indeed to be wished, that these histories should be continued on the plan which Dr. Priestley has adopted. So that all the prominent facts should be collected in the order of their discovery, and a full view be given of the ground already gone over. Abridgments, do not answer this purpose; the theories that dictated the experiments are not detailed, their truth or their fallacy cannot be judged of, and sufficient merit is not attributed to the labours of the discoverer, or the bearings of his facts on his theory, sufficiently explained. To attain gradually to the summit of the temple of science, we must not only build on the foundations of our predecessors, but know somewhat of their intentions at the time of laying them.

The minor treatises of Dr. Priestley on electricity, perspective and natural philosophy, have this discrimination of character, that they are more calculated to allure young people to the study of those subjects than almost any of the introductions which have
either

either preceded or succeeded.   Philosophy is made, not an abstruse science, but a delightful amusement. Indeed it was the fort of Dr. Priestley to make knowledge intelligible and popular, and treat it in such a way, as to invite rather than deter, those who were inclined to enter upon these delightful pursuits. The plainness and simplicity of his syllabus, the amusing complexion of the Phenomena, by which he illustrates his doctrines, and the facility with which the one can be made, and the other compre-hended, affords a very useful example to those who may have the same object hereafter in view.   This was doubtless, owing to his long experience as a teach-er :   and his success in that capacity among his pu-pils, with the electrical machine, and the air pump, is full evidence of the practical utility of his plans of instruction.

Catalogue

T

Catalogue of Dr. Priestley's smaller pamphlets and uncollected papers on philosophical subjects.

Nicholson's Journal. new series. }

V. 1 p. 181.	Reply to Mr. Cruikshank's.
Ibid 198.	Experiments on the Pile of Volta.
V. 2 p. 233.	On the conversion of iron into steel.
V. 3 p. 52.	On air from finery cinder and charcoal.
V. 4 p. 65.	Farther reply to Mr. Cruikshank's.

Amer. Trans.

V. 4 p. 1.	Experiments and observations relating to the analysis of atmospherical air.
V. 4 p. 11.	Farther experiments relating to the generation of air from water.
Ibid p. 382.	Appendix to the above articles.
Ib. Vol. V. { p. 1.	Experiments on the transmission of acids and other liquors in the form of vapours over several substances in a hot earthen tube.
p. 14.	Experiments on the change of place in different kinds of air through several interposing substances.

Republished

p. 21

21. Experiments relating to the absorption of air by water.

28. Miscellaneous experiments relating to the doctrine of phlogiston.

36. Experiments on the production of air by the freezing of water.

42. Experiments on air exposed to heat in metallic tubes.

together.

New-York Med. Repos.   *Title and Date.*

Vol. 1 p. 221.   Considerations on the doctrine of Phlog. and the Decomp. of water. (Pamphlet) 1796.

Ibid p. 541.   Part 2d of do.   (Pamphlet 1797.)

Vol. 2 p. 48.   (Pamphlet) to Dr. Mitchell.

Ibid p. 163.   (Pamphlet) on Red Precipitate of Mercury as favourable to the doctrine of Phlogiston, July 20, 1798.

Ibid p. 263.   Experiments relating to the calces of metals communicated in a fifth letter to Dr. Mitchell.   October 11, 1798.   (Pamphlet.)

Ibid p. 269.   Of some experiments made with ivory black and also with diamonds. (Pamphlet) 11 October, 1798.

T 2   Ibid. p. 383.

Ibid p. 383.   On the phlogistic theory, January 17,
                  1799.   (Pamphlet.)

Ibid p. 388.   On the same subject.   February 1,
                  1799.

Vol. 3 p. 116.   A reply to his antiphlogistian oppo-
                  nents, No. 1.

Vol. 4 p. 17.   Experiments on the production of
                  air by the freezing of water.

Ibid p. 135.   Experiments on heating Manganese
                  in inflammable air.

Ibid p. 247.   Some observations relating to the
                  sense of hearing.

Vol. 5 p. 32.   Remarks on the work entitled "A
                  brief history of epidemic and pesti-
                  lential diseases," May 4, 1801.

Ibid p. 125.   Some thoughts concerning dreams.

Ibid p. 264.   Miscellaneous observations relating
                  to the doctrine of air, July 30,
                  1801.

Ibid p. 390.   A reply to Mr. Cruikshank's obser-
                  vations in defence of the new system
                  of chemistry, 5 Vol. Nicholson's
                  Journal p. 1, &c.

Vol. 6 p. 24.   Remarks on Mr. Cruikshank's ex-
                                        periments

periments upon finery cinder and charcoal.

APPEN-

APPENDIX, NO. 2.

Of Dr. Priestley's Metaphysical Writings.

THE principal source of objection to Dr. Priestley in England, certainly arose from his being a dissenter; from his opposition to the hierarchy, and to the preposterous alliance, between Church and State: an alliance, by which the contracting parties seem tacitly agreed to support the pretensions of each other, the one to keep the people in religious, and the other in civil bondage. His socinian doctrines in theology, and the heterodoxy of his metaphysical opinions, though they added much to the popular outcry raised against him, were not less obnoxious to the generality of Dissenters, than to the Clergy of the Church of England. Nor is it a slight proof of the integrity of his character, and his boldness in the pursuit of truth, that he did not hesitate to step forward the avowed advocate of opinions, which his intimate and most valuable friends, and the many who looked up to him as the ornament of the dissenting interest, regarded with sentiments of horror,

ror, as equally destructive of civil society and true religion.

The extreme difference observable between the apparent properties of animal and inanimate matter, easily led to the opinion of something more as necessary to thought, and the phenomena of mind, than mere juxta position of the elements, whereof our bodies are composed. The very antient opinion also of a state of existence after death, prevalent in the most uncivilized as well as enlightened states of society, confirmed this opinion of a separate and immortal part of the human system : for it was sufficiently evident, that no satisfactory hopes of a futurity after death, could be founded on the perishable basis of the human body. It is only of late days, and from the extention of anatomical and physiological knowledge, that the theory, and the facts of animal organization have been at all understood ; and without the conjunction of physiology with metaphysics, the latter would have remained to eternity, as it has continued for ages, a mere collection of sophisms, and a science of grammatical quibbling. The doctrine of a future state, and that of an immaterial and immortal soul, became therefore mutual

T 4   supports

supports to each other; and herein the civil power willingly joined in aid of the dogmas of metaphysical theology, from observing the convenience that might arise in the government of civil societies, from inculcating a more complete sanction of rewards and punishments for actions in this life, by means of the dispensations in a life to come. Other causes also gave an universal preponderance to the theory of the human soul. It became, for the reasons above mentioned, not only a favourite doctrine with churchmen and statesmen, but the self delusions among the vulgar, respecting supposed appearances after death, rendered it also a *popular* doctrine. Indeed, in every age, and in every country, the priesthood have found it so powerful an engine of influence over the minds of the people, and in too many cases, so fruitful a source of lucrative imposture, that its prevalence is not to be wondered at, wherever artificial theology has been engrafted on the simplicity of true religion, and supported by an established clergy. Of Popery, which yet remains the prevailing system of the christian world, it is doubtless the corner stone; and even under every form of ignorant and idolatrous worship through-

out

out the globe, it is the main source of power and profit to that class of society, which regulates the religious opinions, rites and ceremonies of the country. Not that I would insinuate, that the belief of a separate soul, like some other opinions that might be mentioned, has been generally taught by professors who disbelieve it; for plausible arguments are not wanting, to give it that currency which it has so long received among the wisest and the best of men: nor that an established priesthood of any age or country, or of any religion, is a mere compound of fraud and imposture, for I well know that the wise and the good are abundant in this class of society, as well as in others. But even such men are liable to the common infirmities of human nature; they cannot be indifferent to their rank in society, or the means of their subsistence: it is not every college youth, that is able or willing to weigh "the difficulties and discouragements attending the study of the Scriptures," so forcibly pointed out in the melancholy pamphlet of Bishop Hare: nor is it every professor of christianity, who doubts of the doctrines he has undertaken to teach, that has fortitude enough to follow the noble example of Theophilus

philus Lindsey, and John Disney. Hence we may
take for granted, that those opinions will be admitted the most readily, and enforced the most willingly, which contribute to the influence of that order,
which the professors have been induced by choice,
or compelled by necessity, to wed for life. Choice
indeed, at least that kind of choice, which depends
on a well-grounded conviction of the object chosen
being the means of superior usefulness, has little to
do in this business. For though the clergy of the
church of England severally declare that they are
moved by the Holy Ghost to take upon them the
clerical character, is there one among them in the
present day (Bishop Horsely perhaps excepted) who
would venture to defend this declaration in the sense
originally intended ? It is a fact notorious, that the
candidates for holy orders, regard the profession of
Divinity as they would that of Physic or Law, a fair
and reputable means of gaining a livelihood, by performing those duties which are considered as necessary to the well being of society. It is a fact too,
equally notorious, that wherever theological opinions (like that of the human soul) have been fit and
liable to be made subservient to the temporal pro-

fit

fit or influence of the clergy, that use has been so
made of them by the ambitious and designing part of
the profession, and the rights of the people have been
encroached upon, to serve the interest of the Hierarchy. Nor is it the established clergy alone that
some of the preceding remarks will apply to :
much bigotry among the clergy of the dissenting
interest, may fairly be ascribed to similar causes,
though by no means operating in the same degree.

But important as this doctrine is to the clerical order in political societies, some latitude of doubt and
even of denial, has been conceded in England to the
known friends and adherents of the established system in that country. This is the more to be wondered at, as they have generally considered a dissonance of opinion among their own order, more fatal
to the common interest, than the attacks of their a-
vowed enemies. Thus, more notice was taken of
the Arian heterodoxy of Dr. Clarke, than of the a-
vowed infidelity of Collins, Tindal, Toland, Coward, and other writers of that class, who published
about the same period.

The learned Mr. Henry Dodwell as he is usually
called, and who is a pregnant instance that learning
does

does not always persuade good sense to inhabit the same abode, took great pains to shew that the soul was naturally mortal, but might be immortalized by those who had the gift of conferring on it this precious attribute.   This power he ascribed to the Bishops.   Dodwell, though he would not at first join the establishment, changed his opinion and his conduct in this respect afterward. Bishop Sherlock denied that the existence of the soul could be made evident from the light of nature, (Disc. 2 p. 86. disc. 3 p. 114) Of the same opinion was Dr. Law who quotes him. Archbishop Tillotson declares (v. 12 serm. 2.) that he cannot find the doctrine of the immortality of the soul expressly delivered in scripture.   Dr. Warburton wrote his " Divine legation" to prove that Moses and the Jews neither believed in, nor knew of a future state.   Dr. Law, afterward Bishop of Carlisle, in the appendix to the third edition of his " Considerations on the theory of religion," compleatly overthrows the whole doctrine of a separate soul as founded on the scripture, by a critical examination of every text usually adduced in its support.   Dr. Watson the present Bishop of Landaff in the preface to his collection of theological tracts dedicated to young divines

for

for whose use it was compiled, expressly declares that the question respecting the materiality or immateriality of the human soul, ranks among those subjects on which the *academicorum* εποχη may be admitted, without injuring the foundations of religion. It should seem therefore, that it is not heterodoxy in mere speculative points of theology, that constitutes the sin against the holy Ghost with an established clergy, but heterodoxy on the subject of church authority and the grand alliance. It is in this spirit that the then Archdeacon of St. Albans, Dr. Horsely complains of Dr. Priestley's history of the corruptions of christianity. " You will easily conjecture (says " the Archdeacon in his animadversions on that work " p. 5) what has led me to these reflections, is the " extraordinary attempt which has lately been made " to *unsettle the faith and break up the constitution of* " *every ecclesiastical establishment in Christendom,* " Such is the avowed object of a recent publication " which bears the title of a history of the corruptions " of christianity, among which the catholic doctrine " of the trinity holds a principal place."

This is an unfortunate exposure of the cloven foot of Hierarchy. It was not the wish to detect error or

to

to establish truth—it was not from anxiety to fix upon a firm footing, some great and leading principle of
christianity—it was not the benevolent design of communicating useful information on a litigated topic of
speculative theology—it was not the meek and gen-
tle spirit of sincere and patient enquiry that dictated
those animadversions—all these motives would not
only have borne with patience, but would have wel-
comed and exulted in a temperate discussion of unsettled opinions, before the tribunal of the public ;
for by such discussions alone, can the cause of truth
be permanently and essentially promoted.   No :
these were not the motives that influenced the Archdeacon of St. Albans.   It was the nefarious and un-
pardonable attempt to unsettle the faith of established
creeds ; however founded that faith might be, on ignorance or prejudice, on pardonable misapprehen-
sion, or culpable misrepresentation, on fallacy, on
falsehood, or on fraud.   These " Animadversions,"
proceeded from the morbid irritability of an expectant
ecclesiastic ; from a prudent and a prescient indul-
gence of the *esprit de corps* ; from a dread too perhaps, lest the tottering structure of church establishment, with all its envied accompaniments of sees and
                                        benefices,

benefices, of deaconries and archdeaconries, and ca-
nonries, and prebendaries, and all the pomp and pride
of artificial rank, and all the pleasures of temporal
authority, and lucrative sinecure connected with it,
might be too rudely shaken by sectarian attacks.
But enough for the present, respecting these learned
labours of the Archdeacon of St. Albans; which
like those of Archdeacon Travis may well be considered as having sufficiently answered the *main* purpose of their respective authors, in spite of the wicked replies of Priestley and Porson. Let us say with
the public, *requiescant in pace*.

To return however to the more immediate subject
of the present section. Hobbes seems to have been
the first writer of repute (in England at least) who
denied the doctrine of an immaterial and naturally immortal soul. This was a necessary consequence of his
faith being apparently confined to corporeal existence,
an opinion deducible in fact from the old maxim of
the antients and of the schools, *nil unquam fuit in
Intellectu, quod non prius erat in Sensu.* Hobbes's Leviathan was published about 1650 or 1651.
Spinosa who published after Hobbes was rather an
Atheist than a Materialist, a character to which

though

though Hobbes's opinions might lead, he does not assume. In 1678 Blount sent forward to the public his " *Anima Mundi*, or an historical narration of the " opinions of the antients concerning man's soul after this life according to unenlightened nature," which met with much opposition and some persecution; as was likely, for it is by no means destitute of merit.

In 1702 appeared a book entitled " second " thoughts concerning the human soul, demonstrat- " ing the notion of a human soul as believed to be a " spiritual and immortal substance united to a hu- " man, to be an invention of the heathens and not " consonant to the principles of philosophy, reason, " or religion by E. P. or Estibius Philalethes." " The year following a supplement was published " entitled " Farther Thoughts, &c." The author preoccupies a path subsequently taken by Dr. Law and Dr. Priestley, and endeavours to shew at length that the notion of an immaterial, im- mortal soul, is not countenanced by the texts of scrip- ture usually adduced in favour of that opinion. These texts he criticises individually with a reference to the original words used. The author appears in the

the character of a sincere christian. A second edition of this book was published 1704. In 1706 Mr.
Dodwell before mentioned, a learned and laborious
but weak man, and bigotted to the hierarchy, published his " Epistolary discourse proving from the
" scriptures and the first fathers that the soul is a
" principle naturally mortal, but immortalized actu-
" ally by the pleasure of God, to punishment or re-
" ward; by its union with the divine baptismal spi-
" rit. Wherein is proved that none have the pow-
" er of giving this divine immortalizing spirit since
" the apostles, but only the bishops." This gave
rise to the controversy between Clarke and Collins on
the immortality of the soul. Dodwell's book was
attacked by Chishull, Norris and Clarke. He replied in three several publications, 1st. " A prelimi-
" nary defence of the epistolary discourse concern-
" ing the distinction between soul and spirit, 1707.
" 2nd. The scripture account of the eternal rewards
" or punishments of all that hear of the gospel, with-
" out an immortality necessarily resulting from the
" nature of souls themselves that are concerned in
" those rewards and punishments, 1703. 3d. The
" natural mortality of human souls clearly demon-

U   " strated

" strated from the holy scriptures and the concurrent
" testimonies of the primitive writers. 1708.

About this time Toland in his letters to Serena,
(1704) gives an " Essay on the history of the soul's
" immortality among the Heathens," deducing that
doctrine from popular traditions supported by poetical fictions, and at length adopted and defended
among the philosophers. Concluding from hence,
(preface) that divine authority was the surest anchor
of our hope and the best if not the only demonstration of the soul's immortality; an indirect denial of
the whole doctrine as coming from Toland, who was
certainly no friend to christianity and no believer in
the divine authority of the scriptures.

In the same year (1704) but somewhat previous to
Toland, Dr. Coward had published his "Grand
" Essay, or a vindication of reason and religion
" against impostures of philosophy; proving accord-
" ing to those ideas and conceptions of things human
" understanding is capable of forming itself. 1st.
" That the existence of an immaterial substance is a
" philosophic imposture and impossible to be con-
" ceived. 2ndly That all matter has originally cre-
" ated in it, a principle of internal or self motion.
                                                " 3rdly

" 3rdly That matter and motion must be the foun-
" dation of thought in men and brutes." Dodwell
and Toland had learning enough and so had Blount
to throw some light on the history of this question,
and the author of second thoughts has many observations well adapted to the question he discusses, but
very little is to be gained from a perusal of Coward's
book.

Dr. Hartley's great work, (great, not from the
bulk, but the importance of it) was first published
in 1749. The direct and manifest tendency of the
whole of his first volume is to destroy the common
hypothesis of an immaterial soul: and this he does
with a mass of fact and a force of reasoning irresisti-
ble. He shews clearly how all the faculties ascribed
to the soul, thought, reflection, judgement, memory, and all the passions selfish and benevolent, may
be resolved into one simple undeniable law of animal organization, without the necessity of any hypothesis such as that of a separate soul. Yet he does
not appear distinctly to have seen the full weight and
tendency of his own reasoning, and he adopts a theory on the subject, loaded with more difficulties and
absurdities, than even the common hypothesis.

In

In 1757 was published a philosopbieal and scrip-
tural inquiry into the nature and constitution " of
" mankind considered only as rational beings, wherein
" the antient opinion asserting the human soul to be
" an immaterial, immortal and thinking substance
" is found to be quite false and erroneous, and the
" true nature state and manner of existence of the
" power of thinking in mankind is evidently demon-
" strated by reason and the sacred scriptures."
Authore J. R. M. I. Who this author really was
I know not. But from the perusal of his book it is
probable that he was a physician, and had been travelling. The above work he terms the philosophic
or first part, and refers to a longer work of his own
in manuscript which it seems he could not procure
to be published. There is very little new in the
book so far as I could judge.

I do not recollect any other treatise relating to the
subject that excited public attention in England. In
France and Holland La Mettrie began the controversy by his Histoire naturelle de L'Ame, published
at the Hague in 1745 as a translation from the
English of Mr. Charp;* it is a book containing

---

* This is probably one of the inumerable instances of the careless-
ness

many forcible remarks, and did credit to the side of the question which La Mettrie had adopted. Soon after this La Mettrie published L'Homme machine which was burnt in Holland in 1748. This was an honour not due to the formidable character of the work itself, which though it contains some of the common arguments drawn from the physiology and pathology of the human system, is by no means of first rate merit. He whimsically attributes the fierceness of the English, to their eating their meat more raw than other nations. This book was translated and published in London in 1750.

From

---

ness of French authors in quoting English names. La Mettrie most likely meant to ascribe this to Mr. Sharp the Surgeon, with whose reputation he must have been acquainted. I remember Arthur Young Esq. in one of his annals of agriculture complains that a paper of his translated into French was given to Artor Jionge ecuier. Some years ago Mr. Charles Taylor of Manchester (lately secretary to the society of Arts in London) was requested by Lord Hawkesbury to make some experiments to ascertain the value of East India Indigo when compared with the Spanish. Mr. Taylor did ascertain that the former yielded more colour for the same money at the current prices than the latter by above one fourth. In a paper I believe by M. D'Ijonval these experiments are quoted in a note as made by Le Chevalier Charles Tadkos celebre manufacturier de Manchester.

From Mr. Hallet's discoveries the last volume of which was published in 1736 Dr. Priestley has extracted for himself and quoted what he deemed necessary on this question. I do not notice as part of the history of the question Materialism in England, the foreign atheistical publications, such as *Le Systeme de la nature* attributed to Mirabeau the father, *Le vrai sens du Systeme de l'univers* a posthumous work ascribed to Helvetius, *Le Bon Sens* by Meslier, and others whose titles do not now occur to me, because until within these few years, they were hardly known in England, and excited no discussion of the subject there, previous to the work of Dr. Priestley now under consideration.

The Doctor himself says in his preface to the disquisitions on matter and spirit, first published in 1777, that though he had entertained occasional doubts on the intimate union of two substances, so entirely heterogeneous as the Soul and the Body, the objections to the common hypothesis, did not impressively occur to him, until the publication of his treatise against the Scotch Doctors, which was in 1774. Those doubts indeed could hardly avoid occurring to any person who had carefully perused

Hartley's

Hartley's Essay on Man, first published in 1749, and Dr. Law's appendix before mentioned in 1755.

Dr. Hartley has shewn with a weight of fact and argument amounting to demonstration, that all the phenomena of mind, may be accounted for from the known properties and laws of animal organization; and notwithstanding, that for some reason or other he has so far accommodated his work to vulgar prejudice, as to adopt the theory of a separate Soul, though in a very objectionable form, it is evidently a clog upon his system, and unnecessary to any part of his reasoning. Substitute PERCEPTION, and his theory is compleat. Nor indeed is it possible to reject this. Constant concomitance is the sole foundation on which we build our inference of necessary connection : we have *no* evidence of the latter, but the former. Perception manifestly arises from, and accompanies animal organization ; the facts are of perpetual occurrence, and the proof from induction is compleat.

Hartley having laid a sufficient foundation to conclude (as Dr. Priestley has done) that the natural appearances of the human system might be fully explained by means of Perception and As-

sociation

sociation, without the redundant introduction of the
common hypothesis, Dr. Law a few years afterward
compleatly proved to the christian world that though
Life and Immortality were brought to light by the
christian dispensation, the common theory of a separate immaterial and immortal soul, was not necessary to, or countenanced by the christian doctrine.
Dr. Law seems by his preface, to have been fearful
of the consequences of expressing the whole of his
opinion on this abstruse subject, and confines himself in his appendix to the examination of the passages of Scripture usually referred to in favour of the
Soul's immortality. This appendix I believe was
first added to the *third* edition of his Considerations
on the Theory of Religion, published in 1755.

Against Dr. Priestley, any ground of popular
obloquy would be eagerly laid hold of by the Bigots
of the day. The doubts expressed in the examination of Drs. Reid, Oswald, and Beattie, excited so
much obloquy, as to render it necessary for Dr.
Priestley to review his opinions, and renounce or
defend them. The result was, the disquisition on
matter and spirit, the first volume containing a discussion of the question of materialism, the second
that of liberty and necessity.

In

In discussing the former hypothesis, Dr. Priestley denies not only the existence of spirit as having
no relation to extension or space, but also the common definition of matter, as a substance possessing
only the inert properties of extension, and solidity
or impenetrability. The latter he defines in conformity with the more accurate observations of later
physics, a substance possessing the property of extension and the active powers of attraction and re-
pulsion, With Boscovich and Mr. Michell, he
admits of the penetrability of matter, and replies to
the objections that may be drawn from this view of
the subject.

It must be acknowledged that highly curious as
this preliminary disquisition is, it is not only unnecessary to the main argument, but leaves the definition of matter open to the question whether there be
any substratum or subject in which the essential
properties or powers of attracting and repelling in-
here. That these powers really belong to matter,
whatever else matter may be, is evident from the
reflection of light, previous to contact with the reflecting substance and its inflection afterward from
the electric spark, visible along a suspended chain,
from

from the phenomena of the metallic pyrometers,
from the rain drop on a cabbage leaf, &c. And
that matter is permeable, at least to light, is sufficiently evident from every case of tranparency. Still
however it cannot consist of properties alone ; a property must be the property of something. But the
proper and direct train of argument in favour of
materialism is, that every phenomenon from which
the notion of a soul is deduced, is resolveable into
some affection of the brain, perceived. That all
thought, reflection, choice, judgment, memory, the
passions and affections, &c. consist only of ideas
or sensations,(i. e. motions within that organ) perceived at the time. Though, judgment, memory,
being words, denoting different kinds of internal
perceptions, relating only to, and consisting of, ideas
and sensations.* That sensations and ideas themselves,

---

* A *Sensation* is an impression made by some external object on
the Senses ; the motion thus excited is propagated along the appropriate nerve, until it reaches the Sensory in the Brain, snd it is there
and there only, felt or *perceived.*

An *Idea,* is a motion in the Brain, excited there either by the laws
of association to which that organ is subject, or by some accidental
state

selves, arise only in consequence of the impressions of external objects on our senses, which impressions are liable to be recalled afterward by the recurrence of others with which they were originally
associated, agreeably to the necessary and inevitable
law of the animal system. That this is evident in
as much as there can be no ideas peculiar to any of
the senses where there is a want of the necessary
bodily organ, as of hearing, sight, &c. inasmuch as
all these ideas commence with the body, grow with
its growth, and decrease with its decline. That
they can be suspended, altered, destroyed, by artificial means, by accident, by disease. That all these
properties of mind, viz. thought, judgment, memory,
passions, and affections, are as evident in brutes as
in men ; and though the degree be different, it is always accompanied with a proportionate difference
of organization. That perception is clearly the result of organization, being always found with it,
and never without it : as clearly so in other animals

as

---

state of the system in general, or that organ in particular, without the
intervention of an impression on the Senses ab extra as the cause of it.
Such a motion being similar to a sensation formerly excited, and being also felt or perceived is the correspondent *Idea*.

as in the human species; and probably in vegetables though in a still lower degree.* That as all the common phenomena of mind, can be accounted for from the known facts of organized matter without the souls, and as none of them can possibly be attributed to the soul without the body, there is no necessity to recur to any gratuitous theory in addition to the visible corporeal frame. That the doctrine of the soul originated in ignorance, and has been supported by imposture; that it involves gross, contradictions and insuperable difficulties, and is no more countenanced by true religion than by true philosophy.

All this has been shewn with great force of argument and ingenuity by Dr. Priestley in these disquisitions, to which it may safely be affirmed nothing like a satisfactory answer has yet been given, or is ever likely to be given. True metaphysics, like every other branch of philosophy can only be found-

ed

---

* Dr. Percival, Dr. Bell in the Manchester Transactions, and Dr. Watson in the last volume of his essays, have made this opinion highly probable. Many additional observations are to be found in Dr. Darwin's works. I consider it as a theory established.

ed on an accurate observation of facts, and as these become gradually substituted for mere names, our real knowledge will improve.   It is to physiology perhaps that the question of the materiality of the human soul, and even that of liberty and necessity will owe the compleatest elucidation.   Until medical writers brought into view the *facts* relating to animal life, the metaphysical disquisitions on these subjects were involved in an endless confusion of words without precise meaning, and almost always including in their definition a *petitio principii*.   Indeed we are not yet fully apprized either in Law, Physic or Divinity any more than in Metaphysics, that the *species intelligibiles* of the old schoolmen, and the whole class of abstract ideas of the new schoolmen with Locke at their head, are not things, but names.   They are not even either sensations or ideas; they are words, convenient indeed for classification, and used artificially like the signs of Algebra, but they have no archetype.   This is a subject which will probably be better understood ere long by the labours of Mr. Horne Tooke.

Dr. Priestley therefore considered the question of a future state, as now rested on the basis which to

a chris.

a christian is or ought to be perfectly satisfactory; on
the promises and declarations of our Saviour, exemplified by his own resurrection from the dead.   Indeed the circumstances of the whole question of futurity depending on the truth of the christian scriptures and on them alone, is calculated to give them
a peculiar and inestimable value in the eyes of those
who look forward with anxious hope* to a continued

and

---

* There are some persons who do not seem to entertain this anxious
hope.   Mr. Gray the poet seems an instance, from the following passage in his ode Barbaras Ædes aditure mecum (Letters V. 2 p. 44)
though I do not recollect that the sentiment has been noticed before.

Oh ego felix, vice si (nec unquam
Surgerem rursus) simili cadentem
Parca me lenis sineret quieto
Fallere Letho.
Multa flagranti radiisque cincto
Integris, ah quam nihil inviderem,
Cum Dei ardentes medius quadrigas
Sentit Olympus!

I wonder whether Gray ever perused the following lines written
by his friend and Biographer the Revd: Mr. Mason.

Is this the *Bigot's* rant?   Away ye vain!
Your hopes your fears, in doubt, in dulness steep!
Go sooth your souls in sickness, grief, or pain,
With the sad solace of, *eternal sleep.*

Yet

and more perfect state of existence after death. Nor is it of any consequence to the christian, that the manner how this will be effected is not plainly reveal- ed; for it is sufficient that the Being who first gave animation to the human frame, will at his own time and in his own manner for the wisest and best of purposes, again exert the same act of almighty pow- er in favour of the human race, and in fulfillment of his promise through Jesus Christ. Such at least

was

Yet know ye Sceptics, know, the Almighty mind
Who breath'd on man a portion of his fire,
Bad his free soul by earth nor time confin'd
To heav'n, to immortality aspire.
Nor shall the pile of hope his mercy rear'd,
By vain philosophy be e'er destroy'd ;
Eternity ! by all or wish'd or fear'd,
Shall be by all, or suffer'd or enjoy'd.
Mason.

It is still more singular that Dr. Beattie with all his professions of christianity, should not have been aware of the atheistical complexion of the following passage in his " Hermit."

Nor yet for the ravage of winter I mourn,
Kind nature the embryo blossom shall save ;
But when shall spring visit the mouldering urn !
Oh, when shall it dawn on the night of the grave !

was the view of the subject habitually entertained by our author.

Indeed, the natural evidences of a future state were never conceived by any reasonable defender of the doctrine, to be of themselves satisfactory and conclusive.* They were never deemed of more value than to produce a *probable expectation* of a state of future rewards and punishments, and they are certainly contradicted by the known facts relating to the origin, the growth, and decline of the human faculties. Bishop Porteus has collected these arguments, and stated them with as much force as his moderate abilities would permit; but by far the best summary of what has been urged on this as well as on almost every important question of morals and metaphysics, will be found in Mr. Belsham's Elements of the Philosophy of Mind. An excellent compendium, by a gentleman, to whom next to Mr. Lindsey, Dr.

Priestley

---

* Dr. Priestley in his observations on the increase of infidelity published at Northumberland, has a passage which would seem to intimate that a future state might be clearly made out by the light of nature (p. 59, 60) but this is certainly inadvertency, and by no means conformable to his constant, deliberate, sentiments on that subject as expressed particularly in his Institutes.

Priestley appears to have been more attached than to any other.

The SECOND part of the Disquisitions on Matter and Spirit, contains a discussion of the long contested and confused question of Liberty and Necessity.

Dr. Priestley is right in his opinion that this question was not understood by the ancients, nor perhaps before the time of Hobbes. Long ago it appeared to me, that the only writer among the schoolmen who had touched upon it, was Bradwardine in his Book De causà Dei, which I regret that I have no opportunity of consulting here. Many of his observations are extracted by Toplady in his treatise on Liberty and Necessity, and in his life of Zanchius; but Toplady like Edwards, did not completely understand the question ; they connected the doctrine of necessity with all the bigotry of Calvinism.

Hobbes in his Leviathan, and in his reply to Bramhall on liberty and necessity in his Tripos, first truly stated the subject, and shewed that the question was, not whether we can do what we will, but whether the will itself, (i. e. choice, preference, inclination, desire, aversion,) is not inevitably determined by motives not in the power or controul of the agent.

V   Hartley's

Hartley's book, however, shews, or rather leads to the conclusion, that these motives are twofold, *ab extra* and *ab intra*. The action depending on the compound force of the motives ab extra, and the physical state of the animal organs at the moment. For the latter is frequently of itself an immediate cause of voluntary action.

But previous to Dr. Hartley's great work, the question of liberty and necessity had been discussed between Collins and Clark, and Clark and Leibnitz.* Collins's Philosophical inquiry into human liberty, first published in 1715 was the only book on the subject worth reading between the times of Hobbes and Hartley, and a masterly and decisive work it is. This appears to have been translated and repeatedly printed on the continent; Dr. Priestley, who republished it in London, mentioning a second edition in 1756 at Paris, and a third edition when he was

there

---

* I do not find that the controversy about the Soul occasioned by the publications of Blount, Coward, Dodwell, &c. involved the question of Liberty and Necessity, though they touch so nearly. It escaped me a few pages back, that Dr. Coward, was also the author of "Second Thoughts concerning the human Soul." (Estibius Psycalethes) as well as of the Grand Essay.

there in 1774. The controversy was kept alive in Collins's life time by Leibnitz; but he like Dr. Edwards who afterwards wrote in defence of the same side of the question in his treatise on Free will, was too much given to expand his ideas, and obscure the sense by the multiplicity of words which he used to express it. The letters of Theodicèe contain many passages well conceived, but the book is insupportably tedious. Hobbes could condense more argument and information in a page, than would serve Leibnitz for a volume.

To this treatise of Collins, plainly and popularly written, no sufficient answer was or could be given. It must have satisfied the mind of every reader capable of understanding the question, though it omitted to notice many objections which were afterwards taken up and fully answered by Dr. Priestley. Collins in his preface takes pains to have it understood that he writes in defence of *moral* necessity only, and not of *physical* necessity. A distinction without a difference, though taken by all who have succeeded him.

I do not dwell on the controversy between Jackson on the one side in defence of human liberty, and

Gordon

Gordon and Trenchard in Cato's letters, because little was added to the sum of knowledge, on either
side. Jackson had learning and industry, but he did
not understand the question, and had no pretensions
to that species of distinguishing acuteness, so necessary to a good metaphysician.

Dr. Priestley, following the enlarged and cheering
views of the future happiness of all mankind, first
connected by Hartley with this question, shews completely that the doctrine under consideration has nothing to do with the strict calvinistic hypothesis.
That it is sufficiently conformable to popular opinion. That it is the only practical doctrine which in fact
is, or indeed can be acted upon with respect to the
application of reasoning and argument, reward and
punishment. That the formation of character and
disposition, the actual inferences we make from, and
the dependence we place upon them, rest entirely on
the truth of this opinion. That from the nature of
cause and effect, every volition must be the necessary
result of previous circumstances. That the *scientia
contingentium*, the great and insuperable difficulty of
God's pretended for knowledge of uncertain events,
can on no other hypothesis be avoided, and that the
doctrine

doctrine of necessity is perfectly consistent with the great plan of divine benevolence, in the present state, and future destination, of the human race.

These subjects called forth remarks by Dr. Price, Mr. Palmer, Mr. Bryant, Dr. Kenrick, Mr. White-head, Dr. Horseley and others; to all of whom, answers were given by Dr. Priestley.

The controversy with Dr. Price is a pleasing specimen of the manner in which an important subject can be amicably discussed between two friends, and made interesting too, by the manner as well as the matter, without any thing of that " seasoning of controversy" which Dr. Horsely afterward thought so necessary to keep alive the public attention, and which he strews over his polemics with so unsparing a hand. The Bishop had not yet however adopted that stile of arrogance by which he has since been so disgracefully distinguished; and it is to be regretted for the sake of his own character as a gentleman and as a writer, that he adopted it at all. Dr. Horsely should recollect, that those who emulate the insolence of Warburton ought at least to give proofs of equal learning and acuteness; and that bigotry and intolerance in defence of opinions which, though a man may

profess

profess to believe, he can hardly profess to understand, will do no credit to his religious, his moral, or his literary character in the present state of knowledge. But character as a writer, may be a secondary consideration, to one who is determined to verify the saying, that godliness is great gain.*

It has been a misfortune to this question, that it has seldom been treated by persons who knew any thing of the organization or physiology of the human frame; and that it has been complicated with all the prejudice arising from the theological tenets of those who opposed the doctrine of necessity. Every physician knows, though metaphysicians know little about it, that the laws which govern the animal machine, are as certain and invariable as those which guide the planetary system, and are as little within the

controul

---

* Dr. Horseley's polemic strictures on Dr. Priestley's writings, exhibit a singular compound of insolence and absurdity. But he is contented, I presume, if he rises in the church, as he sinks in reputation. Some of his opinions are truly diverting. His theory of divine generation by the Father contemplating his own perfections, and his grave suggestion of the three persons of the Godhead meeting together in consultation, stand a fair chance of being noticed by some wicked wit, who may wish to expose the infirmities of orthodoxy real or pretended.

controul of the human being who is subject to them.
Every sensation therefore, and every idea dependent
on, or resulting from the state of the sensory, is the
necessary effect of the laws of organization by which
that state was produced. But we neither have nor
can have any sensation or any idea, but what is so dependent, or but what thus results ; for we can neither
feel nor think without the brain. The words we use
for the Phenomena termed mental, are mere terms of
classification and arrangement of the sensations and
ideas thus produced, and their combinations. Hence
it follows, that all these phenomena depend on the laws
which regulate the animal system, and are the necessary, inevitable result of those laws. The obscurity
which has enveloped this question, has arisen from
want of due attention to that state of mind (or rather
of body) which we call, the will ; and from the power that animals seem to have over the voluntary muscles. But every Physiologist knows that the state of
the system which calls into action the voluntary
muscles, that is, a state of want, desire or inclination, whether to act or to abstain, is the result of previous circumstances to which the animal is exposed ;
and the action of the voluntary muscles, is equally

the

the result of necessary laws, as those of the involuntary.

The great object of terror to the Divines in this question about Necessity, was the consequence resulting, that God is the author of Sin. Many and subtile were the distinctions made upon this subject by the necessarian theologists among the schoolmen, and down to the middle of the seventeenth century. Richard Baxter the peace-maker, in his Christian Directory, his Catholic Theologie and some other works, has briefly reviewed them all, and as usual distinguished upon them so acutely, that what was not quite clear before, he has most effectually obscured. The prevailing opinion, however, seems to have been, not that God permitted the sinful act (for the reply was unanswerable, that God must be considered, as willing that which he does not prevent when he can,) but that God, in the common course of nature as pre-ordained by him, permitted the action itself to come to pass, but not the intention or quo animo of the actor, in which the sin consists ; or as Gale expresses it in the quaint language of the time, it is "God's pre-determinate concurse to the entitative act."

Indeed,

Indeed, I do not see with the orthodox notions then prevalent, how it was possible on the hypothesis of God's foreknowing and pre-ordaining all that comes to pass, to avoid considering God Almighty as the author of Sin; and to feel repugnance toward a system, which makes the deity inflict eternal punishment on a creature, whose actions he might have controuled, and whose existence he could have prevented. Such manifest injustice might be viewed without horror, by the brutal bigotry of Calvin, but the tenets that drew after them such a consequence, could not be adopted without hesitation and regret, by any, but the most thorough going, unfeeling zealot.

Origen's doctrine of Universal Restitution, was first advanced in England (so far as I know) by Rust, Bishop of Dromore, and Jeremy White, who I believe had been Chaplain to Cromwell. Since that, the labours of Stonehouse, Petitpierre, Newton, Winchester, Chauncey and Simpson, have furnished ground enough for us to adopt it as the doctrine of scripture as well as of common sense. By connecting this doctrine with that of necessity, Dr. Hartley and Dr. Priestley have been enabled to give a full and satis-

factory

factory reply to all the objections that can be drawn
from the theory of necessity, making God the author
of Sin.   Indeed, unless God's foreknowledge be denied, the same difficulty must occur on either scheme :
for he has knowingly and voluntarily adopted a system, in which the existence of evil if not necessary, is
at least undeniable.

Granting the goodness of God, it follows acoord-
ing to Dr. Priestley, that he has adopted that system
which is most conducive to general, and individual
happiness upon the whole ; and that the moral evil
of which for the best purposes he has permitted human
creatures to be guilty, and the physical evil, which
here or hereafter will be the inevitable consequence
of that conduct, are necessary to produce the greatest
sum of good to the system at large, and to each human being individually, considering the situation in
which he has been necessarily placed in respect to the
whole system.   Indeed, moral evil is of no farther
consequence than as it produces physical evil to the
agent, or to others.   And as we see in the system of
inanimate nature, that general good is the result of
partial and temporary evil, and that though the one
follows necessarily from general laws as the result

of

of the other, the good manifestly predomninates, so in the moral system, we have a right from analogy to predict, that good will be the ultimate result of the apparent evil we observe in it : that we shall be the wiser for knowing what is to be avoided ; the better for corrected dispositions ; and that the power, and the wish to receive and communicate happiness, will be enlarged through each successive stage of our existence, by the experience of those that have preceded. So at least thought Dr. Priestley.

Leibnitz states some of these ideas with great force in the following passage, which I am tempted to transcribe entire from his *Essais de Theodicèe ; sur la Bontè de Dieu, la libertè de l' homme, et l'origine du mal,* first published in 1710. (Prem. partie Sec. 7, 8, 9.)*

Accord-

---

* *Dieu* est *la premiere Raison des choses :* car celles qui sont bornèes, comme tout ce que nous voyons et experimentons, sont contingentes, & n'ont rien en elles qui rende leur existence necessaire ; ètant manifeste que le tems, l'espace & la matière unies & uniformes en elles-mèmes, & indifferentes à tout, pouvoient recevoir de tout autres mouvemens & figures, & dans un autre ordre. Il faut donc chercher *la raison de l' existence du monde,* qui est l'assemblage entier des choses *contingentes :* & il faut la chercher dans *la substance qui porte la raison*

According to this opinion of Leibnitz, the operative
motive in the choice of the present system being the
attribute of Benevolence in the Almighty, the existence

---

raison de son existence avec elle, & laquelle par consequent est *necessaire* & éternelle. Il faut aussi que cette cause soit *intelligente :* car
ce Monde qui existe étant contingent, & une infinité d'autres Mondes
étant également possibles & également prétendans à l'existence,
pour ainsi dire, aussi bien que lui, il faut que la cause du monde ait
eu égard ou relation à tous ces Mondes possibles pour en déterm aer
un. Et cet égard ou rapport d'une substance existante à de simples
possibilités, ne peut etre autre chose que *l'entendement* qui en a les
idées ; & en déterminer une, ne peut etre autre chose que l'acte de
la volonté qui choisit. Et c'est *la puissance* de cette substance qui en
rend la volonté efficace. La puissance va à l'*etre,* la sagesse ou l'entendement *au vrai,* & la volonté *au bien.* Et cette cause intelligente
doit etre infinie de toutes les manieres, & absolument parfaite *en puissance,* en *sagesse* & en *bonté,* puisqu'elle va à tout ce qui est possible.
Et comme tout est lié, il n'y a pas lieu d'en admettre plus d'*une.* Son
entendement est la source des *essences,* & sa volonté est l'origine des
existances. Voilà en peu de mots la preuve d'un Dieu unique avec
ses perfections, & par lui l'origine des choses.

8. Or cette suprême sagesse jointe à une bonté qui n'est pas moins
infinie qu'elle, n'a pu manquer de choisir le meilleur. Car comme un
moindre mal est une espece de bien ; de même un moindre bien est
une espece de mal, s'il fait obstacle à un bien plus grand : & il y auroit quelque chose à corriger dans les actions de Dieu, s'il y avoit
moyen de mieux faire. Et comme dans les Mathématiques, quand il

n'y

ence of all that we term *evil*, is with respect to him, and his preordination of it, *good*; for the whole intention and motive of its permission is founded in perfect goodness

---

n'y a point de *maximum* ni de *minimum*, rien enfin de distingué, tout se fait également; ou quand cela ne se peut, il ne se fait rien du tout; on peut dire de même en matière de parfaite sagesse, qui n'est pas moins reglée que les Mathematiques, que s'il n'y avoit pas le meilleur (*optimum*) parmi tous les Mondes possibles, Dieu n'en auroit produit aucun. J'appelle *Monde* toute la suite & toute la collection de toutes les choses existantes, afin qu'on ne dise point que plusieurs Mondes pouvoient exister en differens temps & differens lieux. Car il faudroit les compter tous ensemble pour un Monde, ou si vous voulez pour un *Univers*. Et quand on rempliroit tous les tems & tous les lieux; il demeure toujours vrai qu'on les auroit pu remplir d'une infinité de manieres, & qu'il y a une infinité de Mondes possibles, dont il faut que Dieu ait choisi le meilleur; puisqu'il ne fait rien sans agir suivant la suprême Raison.

9. Quelque adversaire ne pouvant répondre à cet argument, répondra peut-être à la conclusion par un argument contraire, en disant que le Monde auroit pu être sans le peché & sans les souffrances : mais je nie qu'alors il auroit été *meilleur*. Car il faut savoir que tout est *lié* dans chacun des mondes possibles : l'Univers, quel qu'il puisse être, est tout d'une pièce, comme un Ocean ; le moindre mouvement y étend son effet à quelque distance que ce soit, quoique cet effet devienne moins sensible à proportion de la distance, de sorte que Dieu y a tout reglé par avance une fois pour toutes, ayant prévu les prières, les bonnes & les mauvaises actions, & tout le reste ; & chaque

chose

goodness guided by perfect wisdom.   With respect
to the finite beings, by whom evil is permitted to
take place, there can be no doubt on this scheme, but
the balance of existence will be happiness even to
them, whenever by proper discipline they are fitted
to enjoy it.   Perhaps it may be doubted withont in-
fringing on the reverence due to the supreme disposer of all events, whether it would be consistent with
his justice, knowingly and voluntarily to bring into
existence, a sentient being, destined to be permanently miserable.

The question of Materialism, has been discussed
since the disquisition of Dr. Priestley, by Mr. Cooper, who adopts the same side.   Dr. Ferriar of Manchester, has rendered it dubious how far the sentient
principle ought to be confined to the brain, though
the facts he adduces, apply with equal force against
the

---

chose a contribué *idéalement* avant son existence a la resolution qui a
été prise sur l'existence de toutes les choses.   De sorte que rien ne
peut etre changé dans l'Univers (non plus que dans un nombre) sauf
son essence, ou si vous voulez, sauf son *individualité numerique*.   Ainsi, si le moindre mal qui arrive dans le Monde y manquoit, ce ne seroit
plus ce Monde ; qui tout compteé, tout rabattu, a été trouvé le meilleur par le Créateur qui l'a choisi.

the common hypothesis of a separate soul, acting by means of the body. The doctrine of Necessity has been opposed by Dr. Gregory of Edinburgh, but with a weakness of argument, and a petulance of language, that places his work in the lowest rank among the writers who have adopted the same side of the question. It hardly deserved the notice of so good an advocate as Dr. Crombie, who has been the latest author on the subject.

Indeed, the question must now be considered as settled ; for those who can resist Collins's philosophical enquiry, the section of Dr. Hartley on the Mechanism of the mind, and the review of the subject taken by Dr. Priestley and his opponents, are not to be reasoned with. *Interest reipublicæ ut denique sit finis litium*, is a maxim of technical law. It will apply equally to the republic of letters ; and the time seems to have arrived, when the separate existence of the human soul, the freedom of the will, and the eternal duration of future punishment, like the doctrines of the Trinity, and Transubstantiation, may be regarded as no longer entitled to public discussion.

It is for this reason that I have paid no attention to the hypothesis of the Scotch Doctors, Reid, Beattie

and

and Oswald, and have given no detailed account of
Dr. Priestley's examination of their writings. In
deed the perfect oblivion into which these writers
have fallen, and the utter insufficiency of such
young gentlemen and lady's philosophy as they have
adopted, has secured them from further animadver-
sion. The facility with which ignorance can refer
all difficulties relating to the phenomena of mind, to
instinctive principles and common sense, might answer the purpose of popular declamation for a while,
but it could not last; and these writers have fallen
into merited obscurity, notwithstanding the national
prejudice in favour of each other, so prevalent among
the Literati of North Britain.

Some passages in Dr. Reid, however ought to
exempt him from the contempt which is due to the
common system advanced by him and his coadjutors : and his last book on the Active powers of man,
is a work of undeniable merit on a very important
subject, which has not yet been discussed with half
the labour it so eminently deserves. The Synthesis
and Analysis of our ideas, the history and process of
their formation, and the detail of facts attending and
connected with their rise and progress, is comparatively

tively a new subject. Des Cartes, Buffier and
Condillac among the French, Locke, Berkeley
and Hartley among the English, and Hume,
Reid, and Adam Smith among the Scotch, are
almost the only authors worth notice who have
treated it expressly, and most of them only partial-
ly.* Something may be found to the purpose in
Hobbes, and in the first part of Dr. Priestley's examination of Reid, Oswald and Beattie, and more
in the first volume of Zoonomia, § 14 and 15.†
The common sense of Dr. Reid and Co. seems to have
been employed as the *clavis universalis on* this subject by Buffier, in his "First Truths." Hutcheson's
theory of the Moral Sense hardly merits notice, nor
does that of Dr. Price promise to add much to the
stock of real knowledge. We have had enough (*sat
superque*) of occult principles, innate principles, and
instinc-

---

* Dr. Dugal Stewart in Scotland, and the Revd. Mr. Belsham in England, have published Elements of the Philosophy of the mind, the first
inclining to the Scotch School of Metaphysics, the latter to the System
of Hartley ; both of them of merit in their way, particularly (as I think)
that of Mr. Belsham.

† I cannot help thinking Dr. Darwin's obligations to Dr. Hartley
and Dr. Brown ought to have dictated more acknowledgement than
he has condescended to make.

W

instinctive principles, which illustrate nothing, but the ignorance of those who employ them.

For my own part, I am persuaded that no Theory of the mind can be satisfactory, which is not founded on the history of the Body. I know of no legitimate passport to Metaphysics but Physiology. Hence I cannot estimate highly the writings of the Scotch Metaphysicians. There is one other feature also common to this School, which satisfies me of their incompetence to this subject; their slight notice, and ambiguous approbation of a man so superior as Dr. Hartley, and their utter ignorance or neglect of the theory he has advanced. On every subject relating to the phenomena of mind, Dr. Hartley's book must be adopted as the ground work of the reasoning, or his principles must be previously and distinctly confuted.*

There

---

* Dr. Reid in his last work has given a critique on Dr. Hartley's theory without understanding it, or even touching on the important points. That theory in substance is this : an external object (a peach for instance) makes an impression at once, on our organs of feeling, of sight, and of taste. The impression thus made on the extreme end of the appropriate nerve, is propagated by some species of motion along

the

The Metaphysics of the present day require also, a more accurate attention to the Theory of Grammar than has hitherto been paid by writers on the sub-

ject.

the course of the nerve up to the brain, and there, and there only, perceived; for if the nerve be cut, or tied, or palsied, in any part of its course, the impression is not perceived. Motions in the brain thus produced, and perceived, are *sensations :* similar motions arising, or produced without the impression of an external object, are *ideas.* These impressions being in the instance given, simultaneous or nearly so, are associated, so that the sensation produced by the sight of a peach, will give rise to motions in the brain similar to those produced at first by the taste and the touch of it : i. e. it will suggest the *ideas* of taste and touch, and excite the inclination to reach and to eat the object of them. Hence sensations, ideas, and muscular motions are associated together and mutually suggest and give rise to each other. What species of motion it is, with which the nervous system is affected in this process, or whether Sir Isaac Newton's Æther, or its modern substitute the electric fluid, has any thing to do with it or not, is no essential part of the theory, and may be adopted or rejected without prejudice to the main system. Some kind of motion there manifestly is ; I think it *demonstrable* that it is vibratory ; but of whatever kind it be, its existence in the brain is unquestionable ; and the association and catenation of individual motions in the brain according to certain laws, is equally so. This is matter of fact, and it was Dr. Reid's business if he could, to shew that neither the motions, the perceptions, or the associations took place in that organ. The general law is expressed by Hartley Prop. 20. Cor. 7.

ject. Perhaps I do not assert too much in saying that we have had no grammarians worth notice, none who have thrown light on the principles of Grammar, but Locke and Horne Tooke. What dreadful confusion has arisen from treating words denoting what are called abstract ideas, as if they were the exponents of real individual existence ? Whereas they are merely signs of artificial classification without any individual archetype. For instance in relation to the present subject, what volumes of laboured and learned trifling have been written on the *Will,* the *Judgment,* the *Understanding* and the other faculties as they are called, of the soul! Yet nothing is more certain than that the will, the judgment, the understanding, &c. have no existence : they are words only, the counters employed in reasoning, convenient signs of arrangement, like the *plus* the *minus* and the *unknown quantity* in Algebra, but no more. The time however is approaching, when Metaphysics will take rank among the Sciences that lay claim, if not to absolute demonstration, yet to an approximation to certainty sufficient for all the purposes of ethical reasoning, and all the practical duties of human life.

APPENDIX, NO. 3.

Of Dr. Priestley's Political Works and Opinions.

DR. PRIESTLEY'S literary character may be viewed in the different lights of a natural philosopher, of a metaphysician, of an ecclesiastical historian, of a defender of religion in general, and of unitarianism in particular, and as an author in the wide field of miscellaneous literature. But there is another aspect in which he may be considered; the result of a few pages indeed, but of equal importance in my opinion with any or with all of these, viz. as a writer on the theory of politics : a subject in which the developement of a simple truth in such a manner as to impress it on the mind of the public, may influence to a boundless extent the happiness of millions. I well know the obloquy and the sarcasm attached to political reformers, and I am ready to acknowledge, it is possible that the melancholy theories of the present day, which judge of the future lot of mankind upon earth, from the history of past facts, may be too well founded ; that war, pestilence and famine, and vice and misery in all its hideous forms may be necessary to

counter-

counteract the over increase of the human species,
and make up for the difference between the arithmetical progression of subsistence, and the geometrical
ratio of accumulating population.* Still the philosopher will have motives to labour in devising methods
for the diminution and the cure of moral and physi-

cal

---

* The objections to the progressive amelioration of the state of
mankind are fully and forcibly stated in that important work of Dr.
Malthus, the Essay on Population.   But I am well persuaded that
much good may be brought about, without danger of too great population, by gradually putting in practice well founded theories of political reform.   I say gradually, for I am no friend to sudden, extensive
and violent innovations.   I wish this very important book of Dr. Malthus were well answered, for I cannot help thinking it will admit of
a reply favorable in a high degree to the schemes of those writers
whom it is written to expose.   Some few ideas I have suggested in
the text, that to me make the prospect more consoling than it would
appear from an implicit confidence in the pictures delineated by his
sombre pencil.

Dr. Darwin (Temp. of Nat. quarto p. 159) has nearly the same
thought with Malthus.

> Human progenies, if unrestrain,d,
>
> By climate friended, and by food sustain'd,
>
> Oer seas and soils, prolific hordes! would spread,
>
> Erelong, and deluge their terraqueous bed :
>
> But war, and pestilence, disease and death,
>
> Sweep the superfluous myriads from the earth.

cal evils, at least as well founded as those of a patient, who reasonably applies the known remedies for
the disease by which he is oppressed. The quantum of evil required to effect the necessary depres-
sion of encreasing numbers, is not yet ascertained;
but it is fully and completely ascertained by the melancholy pages of history for these two thousand
years, that far more evil has been inflicted on the
human race from their ignorance of the means of
preventing it, than would suffice for the purpose:
and that the inhabitants of the earth have been thin-
ned far indeed beyond the required number of comfortable subsistence. What country is, or ever yet
was, so far as we know, so fully populated as not to
be, and to have been, capable of sustaining many
more than ever lived upon a given extent? At what
period of history might not the resource of colonization have been resorted to? When and where has
the theory and the practice of agriculture, and the
economy of produce been perfected? What nation
has not been *depopulated* in its turn, by wars of interest, of ambition, of folly, of ignorance, and of pride?
In what country has not the natural tendency to improvement, and to the support of multitudes been

X 2   kept

kept back, by causes depending solely on the political
ignorance of its inhabitants? Should population be
excessive five hundred years hence, it is fair to presume that the encreased knowledge of the day will be
adequate to the evil; and if not, those who suffer,
will at least be far more competent to the remedy
than we can be. To them let us leave it. At present, the earth does not support above a tenth of the
human creatures that might find subsistence by its
cultivation, and yet we are the daily victims of all the
miseries that flow so plenteously from the wretched
maxims of government to which the nations of the
earth submit.

The arguments of these disconsolate philosophers
may be urged any where, at any time, under any circumstances, with equal propriety. However imperfect the state of any civil community may be, the
reformers are always liable to the objection, that let
them do their best, the evils inevitably attendant on
human nature, will ultimately counteract their efforts.
It is the unanswerable argument of sloth against industry: why take so much trouble for convenience
and comfort, when the same labour will be equally
necessary to morrow as to day in despite of all your
pains?

But

But if the given state of human affairs will obviously admit of improvement, there is a justifiable
motive for a friend of mankind to labour for the
public good.   Is it not evident for instance, that a
greater mass of human happiness might be condensed on the same space, by changing the inhabitants
from a horde of indian savages to a populous and
well regulated community, proportioned in numbers
to the fertility and extent of the territory assigned to
them ? So in the civilized countries of Europe, if
the poor could be better taught, and better fed, and
better cloathed, and better attended in pain and sickness, would not the quantum of human happiness be
increased, even suppose the numbers continued the
same ? If in one state of things, the given term of life
of any individual be 60 years for instance, and the
amount of pain he should endure be expressed by 10,
would not the sum of misery be lessened by lessening
the amount of pain 5 or 6 degrees out of the 10 ?
Yet the dreadful mischiefs of superabundant population would not be increased one jot by such an operation.   The best cultivated countries upon earth
have not yet arrived at their maximum of population.
Of Great-Britain at least a third is uncultivated. and of

the parts under actual cultivation a very small proportion indeed is so well managed, as to exclude future improvement : what a difference between the
four crops a year of the gardener, and the single crop
of the farmer ? It is by no means ascertained either,
what produce is the best calculated to afford the
greatest nutriment, conjointly with the most pleasura-
ble sensations when taken as food. When we have
obtained the produce, the art of cookery is yet in its
infancy, and the same quantity may be made to go
much farther as a pabulum to the human frame, than
the present state of culinary practice will admit of.
Let all these improvements be exhausted, still a well
regulated system of gradual *colonization* is a resource
competent to the wants of future centuries; and
should that fail, some obstacles to the facility of marriage, and some restriction to the numbers of offspring
by milder means than exposure like the Chinese, or
infanticide like the Lacedemonian practice, might
furnish an effectual remedy to any extent. So that the
way is not difficult to be traced by which the bugbear
of overpopulation may be counteracted by less violent and abominable remedies than are usually applied by the tyrants of the earth. We may effect in societies

cieties what we aim to effect among individuals :
Sickness is an evil, but we have already in many instances lessened its pain, its duration and its danger :
Death is an evil, but knowledge and foresight may in
many cases introduce it without pain, as the result
of natural decay instead of the physical misery attendant upon our existence, so often and so unnecessari-
ly suffered by myriads of the human race.   In like
manner may the evil of overpopulation be counteracted, without the necessary recurrence either to
vice or to misery ; and without the dreadful instru-
mentality of political despotism.

If the evils we endure are necessary parts of the
system of nature, the remedies of which we are
permitted to be apprized, are necessary parts of the
same system ; for the one and the other are equally
embraced within its plan.   If we see from the experience of ourselves and others, and if we are taught
by the general tenor of history, that misery is the result of ignorance, knowledge is the obvious remedy ; and we have good reason *a priori* to believe it
will be effectual, or the gradual means of acquiring
and increasing it, would not be placed within our
reach.   Wretched as the present state of civil socie-

X 4

ty

ty is in many respects, no man conversant with the facts of past times, can doubt, but that the state of society in Europe four of five centuries ago was still worse. The dispositions of the mass of mankind were more ferocious, their manners more untamed, the comforts of life more rare, and the sources of pleasureable intercourse, and mutual improvement much fewer than at present. All the good that has been done, has been the fruit of increased knowledge, and there evidently is great room for present and future improvement in spite of the modern despondency of political economists; and though perfection be not attainable, we can as yet set no bounds to approximation: nor are we warranted in believing that any well aimed endeavour to ameliorate the condition of society will be entirely lost. Enough still remains to animate the philanthropist: let us fight with the evils of our own day, and leave posterity to follow the example we set, and maintain the combat until hope forsakes them.

The doctrines of the perfectibility of the species, or at least its continually encreasing tendency to improvement, and to happiness, which Franklin and
Price,

Price, and Condorcet and Godwin have lately supported, was advanced prior to their intimations of
this cheering theory, by Dr. Priestley in the outset
of his treatise on civil government, first published
in 1768, and I shall quote the passage that gave rise
to the preceeding observations.

" Man derives two capital advantages from the
superiority of his intellectual powers. The first is,
that, as an individual, he possesses a certain comprehension of mind, whereby he contemplates and
enjoys the past and the future, as well as the present.
This comprehension is enlarged with the experience
of every day ; and by this means the happiness of
man, as he advances in intellect, is continually less
dependent on temporary circumstances and sensations."

" The next advantage resulting from the same principle, and which is, in many respects, both the cause
and effect of the former, is, that the human species
itself is capable of a similar and unbounded improvement; whereby mankind in a later age are
greatly superior to mankind in a former age, the individuals being taken at the same time of life. Of
this progress of the species, brute animals are more
incapa-

incapable than they are of that relating to individuals. No horse of this age seems to have any advantage over other horses of former ages; and if there
be any improvement in the species, it is owing to our
manner of breeding and training them. But a man
at this time, who has been well educated, in an improved christian country, is a being possessed of
much greater power, to be, and to make, happy, than
a person of the same age, in the same, or any other
country, some centuries ago. And, for this reason,
I make no doubt, that a person some centuries hence
will, at the same age, be as much superior to us."

" The great instrument in the hand of divine providence, of this progress of the species towards perfection, is *society*, and consequently *government*. In
a state of nature the powers of any individual are dis-
sipated by an attention to a multiplicity of objects.
The employments of all are similar. From generation to generation every man does the same that every other does, or has done, and no person begins
where another ends ; at least, general improvements
are exceedingly slow, and uncertain. This we see
exemplified in all barbarous nations, and especially
in countries thinly inhabited, where the connections

of

the people are slight, and consequently society and government very imperfect; and it may be seen more particularly in North America, and Greenland: Whereas a state of more perfect society admits of a proper distribution and division of the objects of human attention. In such a state, men are connected with and subservient to one another; so that, while one man confines himself to one single object, another may give the same undivided attention to another object."

" Thus the powers of all have their full effect; and hence arise improvements in all the conveniences of life, and in every branch of knowledge. In this state of things, it requires but a few years to comprehend the whole preceding progress of any one art or science; and the rest of a man's life, in which his faculties are the most perfect, may be given to the extension of it. If, by this means, one art or science should grow too large for an easy comprehension, in a moderate space of time, a commodious subdivision will be made. Thus all knowledge will be subdivided and extended; and *knowledge* as Lord *Bacon* observes, being *power*, the human powers will, in fact, be enlarged: nature, including both its materials,

and

and its laws, will be more at our command; men will make their situation in this world abundantly more easy and comfortable; they will probably pro-long their existence in it, and will grow daily more happy, each in himself, and more able (and, I believe, more disposed) to communicate happiness to others. Thus, whatever was the beginning of this world, the end will be glorious and paradisaical, beyond what our imaginations can now conceive. Extravagant as some may suppose these views to be, I think I could show them to be fairly suggested by the true theory of human nature, and to arise from the natural course of human affairs. But for the present, I wave this subject, the contemplation of which always makes me happy."

Under these impressions Dr. Priestley sat down to investigate the principles on which governments *ought* to be founded, and by which their claims to public support and approbation ought to be tried.

Many works had been written (in England parti-cularly) in favour of those forms and principles of go-vernment, that might operate as a check on the na-tural tendency of all monarchies to despotism, and on the inevitable encroachments of intrusted power.

The

The old writers on the English constitution Bracton
and Fleta, hold sentiments on the constitutional
rights of the Barons to interfere on occasions of royal misconduct, very hostile to the principles afterwards adopted.

Du Plessis Mornay in the *Vindiciæ contra ty-
rannos* (if that book be his) and Buchanan in his Dialogue *de jure regni apud Scotos* were strong advocates for the right of resistance. These tenets were
supported with still more energy during the discussions that took place in the reign of Charles, 1st.
when the speeches of the disaffected members in
Parliament, the *Lex Rex*, and the *defensio populi*
against Salmasius, brought the question of implicit
obedience before the mass of the people in Great
Britain, as well as before the literati of Europe.
To these succeeded the writings of Milton, Harrington, and Sydney, of which the last were certainly a more compleat defence of republican government than either those of Milton or Harrington.
Milton's was at best but a half way theory. Sir.
Robert Filmer was too highly honoured by the replies of Sydney and of Locke.

The revolution of 1688, called forth Locke's
famous

famous treatise on Civil Government, which is there considered as a *contract* between the Governors and the Governed : an erroneous notion, for it implies the previous independence of each of the contracting parties, whereas the governors are evidently no more than the agents or servants of the people, and paid for dedicating their time to those objects which the people at large are deeply interested in, but cannot attend to.

The same event produced the discussions between Locke and Hoadley on the one side, and Sherlock on the other. Hoadley was not only a strenuous and able defender of the principles of the revolution, but of the general doctrines of *toleration* in religious matters : a word much in vogue, but which would not have been used by any one who had studied the subject to the bottom. What obligation am I under to my neighbour for tolerating my opinions, if I tolerate his? No part of the question, whether of civil or religious liberty was well understood at that time, and the boldest of the advocates for the principles of that revolution, and the rights of conscience, were but timid defenders of the doctrines, they undertook to support. The parliamentary discussions, threw no light whatever on the rights of the people; they were trammelled and reined in, by the forms of parliamentary proceedings, and the difficulty of making precedent coalesce with principle. Much however was done at that period of discussion, in favour of the people: the great event that produced the controversy, made every man alive to the subject; and the foundation was laid for the more accurate and enlightened ideas of after times.

From that time to the publication of Dr. Priestley on Civil Government, I do not recollect any author of note, but very many excellent observations were from time to time thrown out by the opposition leaders in parliamentary debates. These are well selected by Dr. Burgh, in his political disquisitions, a work of great merit, both in the design and execution; and which has contributed very greatly to open the eyes of the public, to the necessity of a parliamentary reform, and of making the pretended representation of the people in the lower house of parliament more efficient, and more truly what it now so falsely imports to be.

In the year 1768, about eight years before the assertion of American Independence Dr. Priestley published

lished his short " Essay on the first principles of civil
government," in which he lays it down as the foundation of his reasoning, that " it must be understood
" whether it be expressed or not, that all people live
" in society for their mutual advantage ; so that the
" good and happiness of the members, that is the
" majority of the members of any state, is the great
" standard by which every thing relating to that
" state must be finally determined.   And though it
" may be supposed, that a body of people may be
" bound by a voluntary resignation of all their rights
" to a single person or to a few, it can never be sup-
" posed that the resignation is obligatory on their
" posterity, because it is manifestly *contrary to the*
" *good of the whole that it shall be so.*"

He divides his subject into *political liberty*, or the
power which the people reserve to themselves of ar-
riving at offices, and *civil liberty*, or the power which
the people reserve over their own actions, free from
the controul of the officers of government.   The
former he considers only (as it really is) in the light
of a safeguard to the latter.

By this general maxim, that no principle of government can be considered as binding if it be manifestly
" contrary

" contrary to the good of the whole," he tests the
expediency of hereditary sovereignty, of hereditary
rank and privilege; of the duration of parliaments,
of the right of voting, with an evident tendency to
those opinions which later experience has sufficiently confirmed; and he expressly declares that " such
" persons whether they be called kings, senators or
" nobles or by whatever names or titles they be dis-
" tinguished, are to all intents and purposes the *ser-*
" *vants of the public,* and accountable to the people
" for the discharge of their respective offices. If
" such magistrates abuse their trust, in the people
" therefore lies the right of deposing and consequent-
" ly of punishing them." (P. 23 of 2nd edit.)

Elsewhere (p. 40) he says, " The sum of what
" hath been advanced upon this head is a maxim than
" which nothing is more true, that every go-
" vernment in its original principles, and antecedent
" to its present form, is an equal republic." These
political principles that do so much credit to the
strength of his mind, and to his foresight, were manifestly the result of his own reflections; for no one
before him that I recollect, had taken up the question on the same ground. The plain and simple

Y   principle

principle which he adopts as the foundation of all his remarks, is so obviously and intelligibly true, that it gives a force and clearness to his reasoning which no other preceding writer* affords an example of. The Jesuits indeed had long before advanced the doctrine that all civil authority was derived from the people, for the purpose of applying the maxim in defence of their own king-killing principles, as appears from the collection of assertions made from their writings in 1757 by order of the parliament of Paris, and from the work of the Jesuit Busenbaum about the middle of the eighteenth century† condemned, a few years before that collection. But this doctrine was advanced by them in such a way as to do no service to mankind, and to bring them and their writings into deserved reproach.

It is to Dr. Priestley then that we owe (so far as my information extends) the first plain, popular, brief and

---

* Dr. Sykes the very able coadjutor of Hoadley, in his answer to the Nonjurors charge of Schism, upon the church of England, adopts a similar principle, but he does not treat the subject in the masterly manner of Dr. Priestley.

† See D'Alembert's account of the destruction of the order of Jesuits in France. Eng. trans. 12mo. p. 27. 139, &c.

and unanswerable book on the principles of civil government; and it has the more merit, as the experiments on government since made in America, had
not then been thought of. The plainness, and simplicity of Paine's reasonings are not so much to be
wondered at, as he had lived for some years in a country, where he had the successful facts under his eye,
where the subject of politics, was the daily and hour-
ly topic of conversation and discussion with man
woman and child, where republican principles were
almost universally adopted in theory, and had been
found effectual in practice on a very large scale.
These observations at least apply to his Rights of
Man; neither do I wish to detract from the great
merit of that admirable writer, either in respect of the
work last mentioned, or his Common Sense; while
society exists, they will be classic books on the theory of government.

Well is it for mankind, and with sincere and heart
felt exultation do I write it, that such books have
been composed and such experiments have been
tried; and honourable is it to the character of this
country, that the grand and simple truths, on which
human happiness so materially depends, were first

seized

seized on, comprehended, and put in force by the whole body of the people here, and that with a steadiness and success, that justifies the fondest hopes of the real friends of man. The political sophisms which despotism has forced upon the human understanding for so many centuries, and which have kept the human race in a state of comparative ignorance and misery, are now seen through; the light of knowledge has gone forth, liable no doubt to be obscured for a time, but hereafter to be extinguished never.

Indeed it was high time to try some new experiment in government; to put in practice some principle different from that which from the beginning of the world had until then been acted upon. From the melancholy page of history we learn that the favorite maxim so steadily adopted and practised by the rulers of the earth, that society was instituted for the sake of the governors, and that the interests of the many were to be postponed to the convenience of the privileged few, has filled the world for these two thousand years at least, with bloodshed, vice and wickedness from one end to the other : while long and melancholy experience has convinced us, that it is the invariable, essential, and natural character of power

whether

whether entrusted or assumed, to exceed its proper limits; and if unrestrained, to divide the world into two casts, the masters and the slaves.

America has begun upon the opposite maxim, that society is instituted not for the governors but the governed ; and that the interests of the few shall in all cases give way to the many : that exclusive and hereditary privileges are useless and dangerous institutions in society, and that entrusted authority, shall be liable to frequent and periodical recals.   It is in America alone, that the sovereignty of the people, is more than a mere theory : is is here that the characteristic of that sovereignty is displayed in written constitutions ; and it is here alone that the principle of federal union among independent nations has been fully understood and practised.   A principle so pregnant with peace and happiness, as Barlow has fully shewn, that it may be regarded as among the grandest of human inventions.   I throw out of consideration the antient as well as the modern communities ignorantly called republics, and I count nothing upon the federalism of the Grecian league. There has been no republic antient or modern until the American.   There has been no federal union on

broad and general principles well understood and digested, until the American union. To a person
conversant in antient history, and in the constitutions
of this country, there is no need of any attempt to
prove these positions. The guiding principle, that
pervades every republic upon this continent, is that
which Dr. Priestley has so happily adopted and so
well explained, *the interest or good of the majority of
the individuals composing each political community.*

After Dr. Priestley's work, the American war
broke out, which gave rise to Dr. Price's tract on
Civil Liberty, well meant and tolerably executed,
but not carrying with it that simplicity, and conviction which attends the work of Dr. Priestley. I do
not recollect any treatise published in England on
the *principles* of government from that time, until
a pamphlet of Dr. Northcote's, which attracted but
little attention, though it had some merit. In America, the *Common Sense* and the *Crisis* of Paine, produced their full effect; but they were little read in
England, or in the other parts of Europe. From
thence until the French Revolution, nothing of moment appeared on the subject, unless we notice the
commentary of the younger Mirabeau on the pam-
                                                      phlet

phlet of Ædanus Burke against the order of Cincin-
nati, the well known dialogue of Sir W. Jones,
between a scholar and a peasant, and a short paper
in the Manchester transactions on the principles of
government, read in that society in 1787, and since
republished with Cooper's reply to Burke.*

The

---

* Perhaps I ought not to have omitted the *Vindication of Natural
Society* generally attributed, and I believe without dispute to Mr.
Burke. This very eloquent and ingenious imitation of the stile of
Lord Bolingbroke, whatever the prefatory pretences may be, carries
within it, full and complete evidence that the author was in earnest
and that the subject is treated *con amore*. It argues the pref rence
of natural over artificial society, on the grounds furnished by the evils
that have afflicted mankind, from monarchical and aristocratical ambition and despotism, and from the bondage we are kept under, by the
Priesthood, and the Law. All these evils are pourtrayed in Mr.
Burke's best manner. He may have been afterward warped by his
interest, and driven to take the side of power by his ambition and his
necessities, but when he penned the Vindication of Natural Society, he
felt as he wrote, or there is no dependence to be placed on internal
evidence. This small but valuable Essay is not inserted in any edition of his acknowledged works that I have heard of. When it was
first published, I know not. The third edition printed for Dodsley is
dated 1780. No collection of Burke's works I believe contains that
fine specimen of indignant eloquence which closes the first volume
of Burgh's political disquisitions, though it is known to be Burke's.

It

The French revolution whose commencement may be dated in 1789, has given rise to a discussion of the great questions relating to the rights of man, which however obscured by the temporary defection of that people, has fixt truth upon a basis too firm to be shaken, and too universal to be confined to one community.* But whatever were Dr. Priestley's theoretical notions of government, he never was an advocate for violent or precipitate reform. Like the generality of the English reformers, he contented himself with wishing in that country, for a more fair and adequate representation of the people in Parliament. His moderation on the subject of change is evident from his published sentiments already quoted p. 135.

To the same purpose is his advice to the students

at

It may be worth while to mention that the late Lord Nugent, a most strenuous opposer of Parliamentary reform, was the author of the " Ode to Mankind" published by Dodsley in his miscellany.

* Among the works thus educed, the Essai sur les privileges, and the L'uesceque le tiers Etat of the Abbé Seyes, and Paine's Rights of Man are certainly the chief. There are some things very finely said on monarchy and hereditary privilege by Godwin, in his political justice, though the book is, in the main, a laboured and injudicious defence of school-boy paradoxes. I have already mentioned the very exellent writings of Barlow.

at the New College at Hackney, in his dedication to the Lectures on experimental philosophy.

" It may not be amiss, in the present state of things, to say something respecting another subject, which now commands universal attention. You cannot but be apprised, that many persons entertain a prejudice against this College, on account of the republican, and, as they choose to call them, the licentious, principles of government, which are supposed to be taught here. Show, then, by your general conversation, and conduct, that you are the friends of peace and good order; and that, whatever may be your opinions with respect to the best form of government for people who have no previous prejudices or habits, you will do every thing in your power for the preservation of that form of it which the generality of your countrymen approve, and under which you live, which is all that can be reasonably expected of any subject. As it is not necessary that every good son should think his parent the wisest and best man in the world, but it is thought sufficient if the son pay due respect and obedience to his parent; so neither is it to be expected that every man should be of opinion that the form of government under which he

happens

happens to be born is the best of all possible forms of government. It is enough that he submit to it, and that he make no attempt to bring about any change, except by fair reasoning, and endeavouring to convince his countrymen, that it is in their power to better their condition in that respect, as well as in any other. Think, therefore, speak, and write, with the greatest freedom on the subject of government, particular or general, as well as on any other that may come before you. It can only be avowed tyranny that would prevent this. But at the same time submit yourselves, and promote submission in others, to that form of government which you find to be most approved, in this country, which at present unquestionably is that by *King, Lords, and Commons.*"

Conformably to these opinions given to others, he remained on his arrival in America, an advocate for moderate reform in the old country, though a decided republican in the new. Nor did he ever become a citizen of the United States, or abjure his allegiance to the King of England, ill as he thought of the measures of government, and of oaths of allegiance of all descriptions. His wishes and his con-

versation

versation always tended to impress the idea, that improvements in each country should gradually progress, according to the respective situations of each, and in conformity to the previous ideas respectively prevalent on the subject of government, among the better informed classes, and the spirit of the times.

In these opinions no friend of mankind will differ from him. If there be any fact better ascertained than another, it is that gradual and peaceable, is in all cases preferable to violent reform. A man may be too wise to do good. His ideas may extend so far beyond the prejudices and comprehension of the day, as to make them appear ridiculous, or to render them impracticable. Utopian, they will be called, according to the proverbial irony applied to Sir T. More's uncommon work of this description. Such theories may have their effect hereafter, but it may be the opposite of wisdom to attempt the practice of them in certain stages of society. On this rock M. Turgot split. This was foreseen and well understood by Dr. Priestley; and it is to the credit of his good sense as well as his moderation, that his advice and example were evidences of his being thus impressed.

Indeed

Indeed his opinions were in some instances, by no means coincident with the fashionable extent of republican doctrines. He was friendly to the Senate of the United States, as being a body more venerable and respectable than the House of Representatives : he favoured though not septennial which he thought too long, yet triennial or biennial elections rather than annual : he preferred the choice of officers to depend rather on electors chosen by the people, than immediately on the people themselves : and he was an advocate for a moderate degree of independence in the representative character ; which he did not approve of being completely under the controul of popular irritation.

It is certainly true that some evils arise from too frequent elections ; but as elections are managed in this country they are far from being troublesome though annual ; certainly less so than if they were triennial. Were electors to be chosen who should chuse the representatives as they do the president, doubtless the ignorance of the community would not be so faithfully represented as it sometimes is on the present plan, particularly in the state governments; but though the experiment may be worth trying,

and

and every day's experience inclines me to think better of it, still I should judge, a priori, that there
would be some danger of the representatives becoming more independent of the people than the good of
the country requires. It certainly is so with the
Senate of the United States, owing to the long period
for which the Senators are chosen. This independence induced me *formerly* to think, that if a sufficient number of representatives were chosen promis-
cuously for the same term to supply both houses, the best Senate (which need not *perhaps* be
more than a second deliberative body) would
be a number chosen for the session, out of the
whole representation, to form another house or Senate ; in which the proceedings of the House of Representatives might be reviewed and rediscussed.
Mankind have had so much of independence among
their governors, that the safest course until we better
know how far we can safely trust, may be to err on the
side of controul. But on these points, we can only judge accurately by means of making the experiment : for government is as much a science of experiment as chemistry ; and it is the business of a
political philosopher to deduce principles from a
                                        close

close observation of, aud a fair deduction from, past facts.

On his political conduct under the administration of Mr. Adams in this country, it is not necessary to say much. Of that administration, weak, wicked, and vindictive, what real republican can speak well? If Dr. Priestley was hostile to it, his opinions coincident with an American majority, were forced from him by the virulence with which he was treated by writers in this country who were more than suspected to be in the pay of the British government. It is enough that whatever he said and did on that subject, has been sanctioned by the American people; and he had the satisfaction to live long enough to see a government whose theory was in his opinion near perfection, administered under the auspices of his friend Mr. Jefferson in a manner that no republican could disapprove. To the end of his days, this was a source of great satisfaction to him, especially as it became more and more evident from the disorders attendant on the French revolution, that if the republican system was required to stand the test of experiment, it was in this country alone, and under such an administration as he witnessed, that it stood

stood any chance of success. At present, the trial justifies the anxious hopes of its supporters, and bids fair to establish beyond all doubt, the superiority of that form of government, on which the political reformers of modern days have rested their most reasonable expectations, and their fondest hopes.

To the first edition of this treatise on civil government were annexed Remarks, on Dr. Brown's proposal for a national code of education : on religious liberty and toleration : and on the progress of civil societies. In the second edition, all these remarks were much enlarged ; and he added also, a paper on the extent of ecclesiastical authority, another on the utility of establishments, and a third containing remarks on some positions of Dr. Balguy on church authority.

Against a national code of education, he argues irresistibly, that the science of education is yet in its infancy ; that the more experiments are made by individuals interested in their success, the sooner will it be brought to perfection ; that the various stations of life require various and corresponding modes of education ; that God and nature have placed children under the controul of their parents for the early
years

years of their lives, and that this parental and filial intercourse is more valuable to the parties than any equivalent that society can bestow; that such a scheme would tend only to perpetuate and impose on posterity the ignorance and prejudices of the rulers of the day : to which he might have added, that such a code of national education embracing a system of principles religious, moral, and political, would be no other than an instrument of ecclesiastical and political tyranny : we should force upon our children the intolerance of the priest, and the tyranny of the statesman, and leave them, mind and body, the tools and the victims of both these species of detestable oppression. That some things may be taught to children in each of these branches of knowledge, as truths to be received and acted upon until they arrive at those years of discretion when they may be able to investigate for themselves, is certainly unavoidable. But it is equally certain, that since positions are received as axioms in one age and country, which are regarded as impositions in another—since there never has been the time in Great-Britain for instance when most of the prevailing opinions on these subjects were not demonstrably false—since there is no position on any one

of

of them that has not been and may not be contested, an honest and judicious parent, will always so state to his children his own opinions, as to leave their understandings in a great degree unfettered, if their education and future prospects should be such as to give them the means of investigating for themselves. During the minority of youth, and ignorance, and inexperience, the sentiments and the knowledge of the parent must be communicated to the child, and become the rule of his opinions and practice ; because they are evidently accompanied to the child with the best and most disinterested evidence that the nature of his situation will permit him to attain. But I have always felt the honesty and the cogency of Locke's observation in some of his posthumous works, that the practice of instilling *right principles* into children, is no more than taking advantage of the ignorance and dependence of their situation ; and imposing on the weakness of their understandings as yet incapable of judging, the errors and prejudices of their instructors, as certain and undeniable truths.

After all the modern publications on education, the science is even yet in its infancy ; nor has the

Z   particular

particular question just now suggested been sufficiently considered, and discussed. One point however seems to me well established, viz. that all interference on the subject on the part of government,
should be confined to furnishing an easy access for
every member of society, to the means of acquiring
knowledge. Public schools supported at public expence, and open to all children, male and female, for
the purpose of learning to read well, to write well, to
attain a knowledge of the principles of Grammar, and
the elements of Arithmetic and Geography, is far e-
nough; it would then be in the power of each member of the state to become competent to all common
functions, and to go further if he have the means and
the inclination. Such a plan would not detract from
the class of labourers, (as Mandeville* would object)
because as to literary attainment, each would start
on terms of equality, and an acquisition common to
all, would raise none above their fellows. I rejoice
that in the state of Pennsylvania, we have a right to
expect a law extending thus far.

The subject of Religious liberty, and Toleration as
it

---

* Essay on Charity, and Charity Schools.

it is called, and the expediency of Church Establishments, are argued by Dr. Priestley, with his
usual force and acuteness; but it is needless to pursue an analysis of his reasoning on questions which
are clearly settled and ought now to be at rest. The
proper object of a magistrate's controul, are *actions*,
not *opinions:* nor can any two things be more distinct than what respects our conduct here, and what
respects our conduct in reference to a future state of
existence. Rulers have forgotten, as Milton observes, that force upon conscience will warrant force
upon any conscience, and therefore upon the con-
sciences of those who now use it. If I tolerate my
neighbour's opinions, and he tolerates mine, we are
upon equal terms; but if he should require me to
renounce my own, and to embrace his, under any
penalty whatever, positive or negative, by the inflic-
tion of actual punishment, or the deprivation of
common privilege, he is obviously and indubitably
a tyrant. I can suggest no argument more plain
and self evident than this. Whether a man believes
in one God with the Unitarians, or in one God and
two thirds with the Arians, or in three Gods with
Dr. Horseley and the Trinitarians, or in thirty or

Z 2   thirty

thirty thousand Gods as Varro tells us the heathens of his day could reckon up, or in no God at all like the Atheists, under any of these modes of belief a man *may* be a good member of society, and under all of them men *have been* good members of society: such a man's course of life may be just and benevolent; he may pay full obedience to the laws; he may be a good father, a good husband, a dutiful son: his *actions*, his *conduct* may be kind, generous and upright: what more has society to require? of what importance are a man's opinions, if his actions are those of an honest man? Is not a life of good conduct with any opinions, better than a life of bad conduct with the most orthodox?* Or of what consequence are good opinions if they do not produce the fruit, of good conduct? can there be better evidence of the orthodoxy of a man's opinions than the uprightness of his conduct? Again; it is absurd to attempt impossibilities: it cannot be the duty of

any

---

* " I have heard frequent use" (said the late Lord Sandwich, in a debate on the Test Laws,) " of the works orthodoxy and heterodoxy, but I confess myself at a loss to know precisely what they mean." Orthodoxy my Lord (said Warburton in a whisper) Orthodoxy, is *my* Doxy: heterodoxy, is *another man's* Doxy.

any man or set of men to make such an attempt:
it cannot then be the duty of a magistrate, or of the
laws to interfere with opinion, because in the nature
of it, it is incontroulable. The man who holds it,
cannot help holding it. His belief, the convictions
of his mind, are the necessary result of the evidence
by which they are produced and accompanied, and
he cannot help having them. All therefore that the
interference of power can effect, is to make him
profess a falsehood, and declare his belief in what
he does not believe: but the opinion itself, can only
be changed, if at all, by reasoning and reflection.

How much more simple then, how much more
practicable is the system, of regulating a man's conduct, and leaving him to regulate his opinions as he
thinks fit. How competent the one is, to all the
good purposes of society, and how productive has
the other been, of vice, of cruelty and misery in every
country upon earth! for to the system of the magi
strate's right to interfere in the regulation of religious
opinions do we owe all the religious wars and perse-
cutions of Pagans against Christians, and Christians
against Pagans, of Papist against Protestant, and
Protestant against Papist——all the proverbial in-

veteracy of that species of rancour which has been denominated (καὶ ἐξοχην) the *odium theologicum.* To this system we owe as in England, the exclusion of good men from offices who will not take a false oath, or sport with a religious ceremony, in order that men who will do both without scruple, may be admitted in their stead: holding out the honours and emoluments of society as the certain rewards of mental dishonesty, and palpable blasphemy. How true is the observation of Dr. Jortin in that inimitable preface to his ecclesiastical history? " Men " will compell others, not to think with them, for " that is impossible; but to say they do, upon which " they obtain full leave, not to think or reason at " all, and this is called Unity: which is somewhat " like the behaviour of the Romans, as it is describ- " ed by a brave country man of ours in Tacitus, " *Ubi solitudinem faciunt, pacem appellant.*"

This question of religious liberty is one of those which the discussions of the last thirty years has brought compleatly within the view of the public. The half way defences of the friends of truth on this subject from Milton to Locke and Hoadley,*

* Perhaps I am wrong in ranking Milton and Locke among the
half

and from thence to Priestley, served to draw some attention to the questions embraced ; but until the essays appeared, which are now under consideration, there had been nothing like a full and free discussion of the subject, nothing that reached *au fond.* Dr. Priestley carried the same mode of reasoning, the same clearness and force, that distinguished his treatise on civil government, into the observations on religious liberty and toleration. We had nothing equal to it before, and I recollect nothing superior since. It is fortunate for mankind, that the experience of this country has come in aid of the doctrines he has advanced, and settled the question by an appeal to fact, in a manner that carries full conviction, and leaves no room for future controversy. America has shewn, that the interests of religion may be sufficiently supported, the peace of society effectually preserved, and the progress of society exist in the most rapid state of improvement hitherto known,

without

half way defenders of religious liberty, a concession that is forced from me by a recollection of the excellent treatise on Liberty of Conscience by Milton, and the still more convincing letters of Locke to Limborch.

Z 4

without the dangerous aid of religious tests, or church establishments, as well as without the needless appendages of bishops, nobles, or kings. Whether the state of knowledge in England would justify any attempt at reformation beyond the long sought object of parliamentary reform, is a question that wise and moderate men may reasonably doubt about, here, all doubts on the subject as connected with the true interest of America, have long vanished; and the people rest satisfied with an experiment which has fixed the theory on a basis too firm to be shaken.*

---

* Dr. Franklin would have had great merit for fabricating that beautiful chapter on toleration so well known and so generally ascribed to him, had he not been a plagiarist in this instance. The passage is to be found in Taylor's Liberty of Prophesying Polem. Discourses fol. p. 1078. The fable however is of Arabic origin as I strongly suspect from the following extract of a dedication to the consuls and senate of Hamburgh of a book whose title is יהודה שבת (Shebeth Jehudah) Tribus Judæ. Salomonis Fil. Virgæ. Complectens varias Calamitates, Martyria, Dispersiones, &c. &c. Judæorum. De Hebræo in Latinum versa a GEORGIO GENTIO, cɔↃ loɔ LXXX (1680)

Dedication p. 3. Illustre tradit nobilissimus autor Sadus, venerandæ antiquitatis exemplum, Abrahamum Patriarcham, hospitalitatis gloria celebratum, vix sibi felix faustumque credidisse hospiti-

hospitium, nisi externum aliquem, tanquam aliquod presidium domi, excepisset hospitem, quem omni officiorum prosequeretur genere. Aliquando cum hospitem domi non haberet, foris eum quæsiturus campestria petit, forte virum quendam senectute gravem, itinere fessum, sub arbore recumbentem conspicit. Quem comiter, exceptum domum hospicem deducit, et omni officio colit. Cum cœnam appositam Abrahamus et familia ejus a precibus auspicaretur, Senex manum ad cibum protendit, nullo religionis aut pietatis auspicio usus. Quo viso Abrahamus eum ita affatur : Mi Senex, vix decet canitiem tuam, sine prævia numinis veneratione, cibum sumere. Ad quæ Senex : ego Ignicola sum ; istius modi morum ignarus ; nostri enim majores nullem talem me docuere pietatem. Ad quam vocem horrescens Abrahamus, rem sibi cum ignicolà pro profano et a sui numinis cultu alieno esse, eum a vestigio a cœnà remotum, ut sui consortii pestem et religionis hostem domo ejecit. Sed ecce summus Deus Abrahamum statim monet. Quid agis Abrahame ? Itane vero fecisse te decuit ? Ego isti seni quantumvis in me usque ingrato et vitam et victum centum amplius annos dedi, tu homini nec unam cœnam dare, unumque eum momentum ferre potes ? Quà divinà voce monitus, Abrahamus senem ex itinere revocatum domum reducit, tantis officiis pietate et ratione colit, ut suo exemplo ad veri numinis cultu eum perduxerit. Vos quoque Proceres nobilissimi cum pari studio Judæorum gentem habeatis, laudatissimo more atque exemplo, pietate potius servare, quam severà disciplinà excludere ; eos tanquam perditas Christi oviculas colligere quam dissipare mavultis.

APPEN-

Of Dr. Priestley's Miscellaneous Writings.

THESE consist principally of his Grammar and Lectures on the Theory of Language, his Lectures on Oratory and Criticism, and those on General History and Civil Policy.

The Grammar was first published in 1761. A month after the second edition of it, Dr. Lowth's Grammar came out. The third and last edition of Dr. Priestley's was in 1772. I do not observe any peculiarity in this work. It seems like all Dr. Priestley's writings and compilations, sensible, plain, and intelligible. Dr. Lowth had at that time more reputation in the world than Dr. Priestley; his lectures de sacrà poesi Hebrœorum, having deservedly procured him the respect of the literary part of the public. His grammar therefore seems to have interfered with the circulation of Dr. Priestley's.

The Lectures on the Theory of Language and Universal Grammar were printed at Warrington in 1762 in one volume duodecimo. I believe though printed and delivered to the students, it was never

fully

fully published;* I shall therefore give an account of the subjects treated in this small work, more at length, than if the treatise itself had been generally known.

The first lecture after the introduction is on *Articulation*, or the power of modulating the voice. This is peculiar, as Dr. Priestley thinks, to the human species. Brute animals, emit sounds, and varieties of sound, the effect and expression of passions and sensations; they have also gestures to make known their wants and feelings : but the superior capability of the organs of speech is perhaps the most important characteristic of humanity. Those articulations he observes are preferred which occasion the least difficulty to the speaker. Very antient languages like the Hebrew, Arabic, Welsh and even the Greek, abound with harsh articulations which are gradually changed.†

Lecture

---

* They are mentioned however with approbation by the writer of all others best able to judge of their merit. See note to Epea Pterocnta 75.

† Dr. Darwin in his notes to the Temple of Nature has some ingenious remarks on the articulation of alphabetical sounds.

Lecture 2d. *On the origin of Letters.* The transition from speaking to writing, is so difficult as to lead some persons like Dr. Hartley to have recourse to supernatural interposition to account for it.* Robertson's Comparison of Alphabets makes it probable that all the known ones have been originally derived from the Hebrew or Samaritan. Dr. Priestley's opinion is that the rude attempts of our earliest forefathers, were improved partly by attention and

---

* Dr. William Scot the Civilian, who was sometime Professor of History at Oxford, in his introductory lecture, urged the impossibility of language itself being originally acquired by human effort, and thence inferred the necessity of recurring to the theory of miraculous interposition. But supposing the still greater difficulty of a man first appearing in a state of manhood, there would be no doubt in my opinion of the gradual acquisition of a collection of significant sounds, if there were another human creature to whom they might be addressed.

Gilbert Wakefield in an "Essay on the origin of Alphabetical Characters" in the second volume of the Manchester Transactions is of opinion that language and alphabet too, are to be attributed in their origin to divine communication, and are not by any means explicable on the theory of gradual improvement. I have no objections to introduce a miracle when we cannot do without it, but I cannot see the Dignus vindice nodus in the present case. Mr. Harvey's Essay on the English Alphabet in the first part of the fourth volume of the same transactions is worth a perusal.

and partly by chance until alphabets were invented. Moreover the imperfection of all alphabets argues that they are not the produce of divine skill: had such a one been revealed, it would certainly have established itself by its manifest excellence.

Lecture 3d. *Of Hieroglyphics, Chinese Characters, and different Alphabets.* Alphabets as they now appear, were not the first attempts at expressing ideas in writing. Picture-writing, or the rough draught of the things meant to be expressed preceded Hieroglyphics which were a contraction of picture-writing. The Chinese letters seem to be a still further contraction of Hieroglyphics.*   All these seem

to

---

* " That there was however a relation between the real Egyptian " Hieroglyphics and the Chinese Characters, De Guignes, so well " versed in the literature of China, undertook to evince ; and actually " composed a work to shew that each of the 214 keys or elements " correspond to Egyptian Hieroglyphics, that they were of the same " shape and signification, and consequently were identified (see M. de " Hauteraye's Alphabets in the French Encyclopædia, and Hist. de " l'Acad. des Inscrip. V. 34.) This work thus announced in 1766, " has never appeared, but remained only a system (as M. de Haute-" raye asserts,) with its author, who died but a few months ago (1801.) " Hager's Ch. Ch. 38 "

There

to have preceded the methodical arrangement of alphabets, In picture-writing, abstract ideas would
be expressed by Metaphors, as eternity by a serpent
biting his tail : impossibility by two feet standing on
water, and so on. The mode of contraction may be
illustrated thus; suppose two swords cross wise represent a battle, two cross strokes may be used in
lieu of the more perfect delineation. Arbitrary characters would also be introduced to express ideas, as
we use the numerals from one to ten. Of arbitrary
characters

There is much curious remarks collected by Dr. Hager in his
magnificent book on the Chinese Character : it seems to me also to
have the merit of being the finest specimen of printing extant. But
Hager's remarks ought to be perused subject to the criticisms of that
very acute and judicious traveller Mr. Barrow. See his travels in
China, Chap. VI.

Dr. Priestley's opinion seems to be the same with Warburton's who
(Div. Leg. L. 4. § 4.) calls the Chinese Character the runninghand of
Hieroglyphics. The Chinese Characters including synonimes are
reckoned at 80,000. A knowledge of 10,000 however, suffices to
read the best books in each Dynasty. Hag. Ch. Ch. 49. The Chinese language is monosyllabic, and consists of 214 keys or elements
and but 350 words. The Japanese (quite unlike it) is polysyllabic,
and contains many more. Ib. 54.

Warburton's Essay on Hieroglyphics is deserving of the character
which Condillac gives it. Essai sur l'orig des Conn. V. 1. Ch. 13.

characters the Chinese writing is said to be full. These have multiplied so exceedingly that it takes a man in that country half his life to learn to read the necessary books, hence improvement is at a stop there.*

The most antient Alphabets are those of the eastern Languages.† The Phenician, Hebrew, and Syriac or Samaritan had the same origin. The derivation of the Greek from these is very evident; the similarity of the letters being easily traced. Cadmus is said to have brought the knowledge of letters from Phenicèa. The order of letters in the Greek Alphabet proves the same thing. The chasms arising from the rejection of such Hebrew letters as the

Greeks

---

* The same remark will apply to the Mexican Hieroglyphics and Characters; for it appears from Clavigero that they had advanced into Characters, and as he thinks as far as the Chinese. But the state of improvement in the two countries, affords no countenance to this opinion. Dr. Hager says there is no similarity between their characters. Ch. Ch. 46. Dr. Priestley's observation is confirmed by ch. VI. of Barrow's travels.

† The Dr. does not seem to have been aware of the Alphabets of Adam, Enoch and Seth, published at Nuremburgh Hersel. Synop. univ. philos. norimb. 1841! Hager's ch. ch. 30.

Greeks had no sounds to, were afterwards filled up by Palamedes and Simonides.*

The Latin was nearly the same with the Greek, before the last additions made to it, retaining the F of the Æolians, and the aspirate H of the Pelasgi. The Greeks denoted all their numbers by the same letters as the Hebrews, and to make their Alphabet tally with the Hebrew for this purpose, they filled

up

---

* The want of alphabet among the Chinese is a curious point of discrimination between them and the other eastern nations. Whether India or China has the highest claims to literary antiquity is not yet fully settled. The following instances of coincidence are as curious as those noticed by Dr. Priestley. " The same division of the Zodiac " among the Greeks and Romans as among the Chinese : the same " number and order of the planets; their application to the same days " of the week as among the Romans are circumstances that could " hardly be accidental." Dr. Hager Ch. Ch. p. XVII. from Mem. des Mission de Pekin V. 1 p. 381.

But coincidence is a dubious ground to rest any theory upon; unless the argument from induction be very full. We may perhaps allow Major Vallancey and Sir Laurence Parsons to have established the identity of the Irish and Carthaginian languages, but the coincidences of Mr. Bryant will not class much higher than those offered between the Welsh and the Greek in some of the early volumes of the monthly Magazine. They are curious and ingenious ; but they lead to no conclusion.

up all the remaining chasms in their old Alphabet with real Hebrew letters. It is further probable that the antient Greeks in imitation of the Phenicians wrote from right to left, and then from left to right, and so on alternately : this method was called βουϛρο- φηδον from its resemblance to plowing : this was before it was fixed in the method in which at length they, and after them all the nations of Europe, have used it, viz. from left to right, without variation. The Chinese and Japanese whose language is not alphabetical, write in neither direction.*

The remaining part of the lecture consists of remarks on vowels and accents, and the history of their

use

---

* The Chinese, Japanese, and Mantchou Tartars write perpendicularly ; de haut en bas. Dr. Hager 57. But the Chinese as well as the Egyptians formerly wrote horizontally as well as perpendicularly. Ib. 45.

The British museum contains two Japanese books in alphabetic letters. Harl. Mss. 7330 and 7331. Hag. 59. The people of Corea also use alphabetic characters. Ib.

Dr. Priestley's observations on the gradual improvement of the hieroglyphic into the alphabetic character, are coincident with those of Dr. Hager; and are verified by the fact, that the most antient Chinese characters are, and are called, images, forms. Ib. 44.

A a

use and application, together with miscellaneous remarks, which though curious and interesting, do not admit of abridgment.

Lecture 4th. *Of the general distribution of words into classes.* In this Lecture Dr. Priestley traces the probable operations of the mind, in distributing and noting nouns, whether of individual things, or of abstract ideas, and adjectives or epithets; thence into articles, verbs, &c. The fine discovery of Horne Tooke, that the class of words usually deemed insignificant of themselves, sre not so, but are in fact resolveable into verbs or participles, or nouns, was not then known.* It were to be wished the Doctor had revised these lectures and made use of the truly original remarks of Mr. Tooke. With Mr. Harris, he considers (p. 142) particles as having no meaning of themselves. Yet in another place he seems to have an idea of the same kind with Mr. Tooke's. "The names of things, or qualities, are
                      "called

---

* Dr. Beddoes seems to think that although Mr. Tooke has full claim to the discovery, something of the general theory has been stated by the Leyden Professors, Hemsterhuls Lennep, Scheid, &c. Observ. on demonstr. Evid. p. 5: And (but subsequently to the Letter to Dunning,) by M. Volvoison.

" called *substantives* and *adjectives*.   The substitutes
" of these are *pronouns*.   Their coincidence or
" agreement is expressed by *verbs*.   The relations of
" words by *prepositions*, and of sentences by *conjunc-*
" *tions*.   ADVERBS *are contracted forms of speech,*
" *which may be analized into words belonging to*
" *other classes*."   Pronouns he considers chiefly as
adjectives.

From the fourth to the ninth Lecture, the remarks
though apparently just and calculated to explain and
illustrate his subject by references to the coincidences and discriminations of other languages, particularly the Hebrew, Greek, Latin and French, are too
technical to be dwelt on in this brief review.

His ninth Lecture is on *adverbs, prepositions, con-*
junctions, &c.   Adverbs he says are chiefly contractions for other words, and often for whole sentences, a
position which the Epea pteroenta has sufficiently confirmed.

He quotes occasionally with implied respect the
Hermes of Mr. Harris ; a book then much in vogue,
and bepraised without stint or consideration by Dr.
Lowth and others.   It may indeed be amusing from
the learned trifling, and strange absurdities where-

A a 2   with

with it abounds; had the author given us a little
good sense in lieu of a great deal of Greek reference, it
would have been better worth reading; but it has now
attained its proper rank among the literature of the
age. According to Mr. Harris, adverbs are attributives of attributives!

The latter part of this Lecture is on *Dialects*, and
contains so ingenious, and to me so novel an account
of their origin, from the circumstances of the dependence or independence of the countries wherein they
obtained, that I am tempted to transcribe the passage.

" When a language was spoken by several independent cities or states, that had no very free communication with one another, and before the use of
letters was so generally diffused as to fix the modes
of it, it was impossible, not only to prevent the
same words being pronounced with different tones
of voice (like the *English* and *Scotch* pronunciation)
but even the number and nature of the syllables
would be greatly altered when the original root remained the same; and even quite different words
would be introduced in different places. And when,
upon the introduction of letters, men endeavoured

to

to express their sounds in writing, they would, of course, write their words with the same varieties in letters. These different modes of speaking and writing a language, originally the same, have obtained the name of DIALECTS, and are most of all conspicuous in the *Greek* tongue, thus εγω *I*, was, by the *Attics*, frequently pronounced εγωγε; by the *Dorians* εγων and εγωνγα; and by the BEOTIANS εωγα and εωγγα."

" All these different modes of speaking, like all other modes, might have grown into disrepute, and, by degrees out of use, giving place to one as a standard, had particular circumstances contributed to recommend and enforce it ; but, in *Greece*, every separate community looking upon itself as in no respect inferior to its neighbours in point of antiquity, dignity, intelligence, or any other qualification ; and being constantly rivals for power, wealth, and influence, would no more submit to receive the laws of language from another than the laws of government : rather, upon the introduction of letters and learning, they would vye with each other in embellishing and recommending their own dialects, and thereby perpetuate those different modes of speech."

A a 3   " On

" On the other hand, in a country where all that spoke the language had one head, all writers, ambitious to draw the attention of the leading men in the state, would studiously throw aside the particular forms of speaking they might happen to have been brought up in, and conform to that of their superiors: by this means dialects, though used in conversation, would hardly ever be introduced into writing; and the written language would be capable of being reduced very nearly to a perfect uniformity."

" For this reason the language of *Greece*, though spoken at first within a very small extent of country, yet by a number of independent cities, had no common standard; and books now extant in it differ so widely in their forms of expression, that the most accurate skill in some of them, will not enable a man thoroughly to understand others. Let any person after reading *Homer* or *Hesiod* take up *Theocritus*. Whereas, in the *Latin*, though written in very distant parts of the vast *Roman* empire, dialects are unknown. However differently *Romans* might speak, there are no more differences in their writings than the different genius, abilities, and views of diffe-

rent

rent men will always occasion. The *Patavinity* of Livy is not to be perceived."

" When a language had been spoken by different nations a considerable time before the introduction of letters and learning, the variations in the forms of speech would grow too considerable to form only different dialects of the same language; when reduced to writing they would form what are called *sister-languages*, analogous in their structure, and having many words in common. Thus the *Hebrew* and *Chaldaic* or *Syriac* with perhaps other eastern languages, might have been originally the same. On the other hand, the *English* and *Scotch*, had the kingdoms continued separate, might have been distinct languages, having two different standards of writing."

The 10th Lecture is *on the Derivation and Composition of words, on Syntax, and on Transition.*

The 11th is *on the Concatenation of Sentences, and the transposition of words.* The following observations among others on the first of these subjects appear to be worth transcribing.

" Now the method of learning and using a language that is formed must be analogous to the me-

thod

thod of its formation at first. Short and unconnected sentences would be sufficient for the most press-
ing and necessary occasions of human life, of men
acquainted with but a few objects, and only the most
obvious qualities of those objects: As human life
improved, as men became acquainted with a greater
variety and multiplicity of objects, and new relations
were percei ed to subsist among them, they would
find themselves under a necessity of inventing new
terms to express them. As their growing experience and observation would furnish them with a
greater stock of ideas to communicate, and subjects
to consult and converse about, their endeavours to
express their new conceptions of things would lead
them, by degrees, and after repeated trials, into every
requisite form of transition, for the purpose of connected discourse, either of the historical, or argu-
mentative kind."

" But, as we find that persons who have not learned
to read and write are in a great measure incapable of
a connected discourse, and even persons who are accustomed to read and converse with ease are seldom
able, at first, to put their thoughts together in writing
with tolerable propriety ; it is not easy to conceive,
that

that the language of any people, before the introduction of letters, should be otherwise than very incoherent and unconnected : and that their first attempts to write would want that variety, accuracy and elegance of contexture, which their late compositions would acquire."

" Hence the striking simplicity of style in the books of the *old testament*; perhaps the most ancient writings in the world. The history of *Moses* how different from that of *Livy*, and *Thucydides*; or even of *Cæsar*, *Sallust*, and *Xenophon*. The moral discourses of *Solomon*, how different from those of *Plato*, *Cicero* and *Seneca*; for though much time had elapsed from *Moses* to *Solomon*; yet the *Hebrew* nation, not having been addicted to letters in that interval, their language had received little or no improvement."

" Even the writers of the *new testament*, having been chiefly conversant with the ancient *Jewish* writers, and their education having given them no leisure to attend to the refinements of style, have generally adopted the simple unconnected style of their forefathers, both in their narration and reasoning. I shall give one instance of this. *John the evangelist*

in

in giving an account of a conversation that passed
between *John the baptist*, and the *Jews;* instead of
carrying on the narration in his own person, as an
historian, and giving the questions and answers such
a form as was proper to make them incorporate with
his own account of them (a turn quite familiar to o-
ther writers) he reports the words just as they were
spoken, notwithstanding the speeches were too short
to make it in the least necessary or expedient to set
down the whole by way of formal dialogue."

"John I. 19. *And this is the record of John, when
the* JEWS *sent Priests and Levites from* JERUSALEM
*to ask him Who art thou? And he confessed, and
denied not; but confessed, I am not the Christ.
And they asked him, What then art thou? Elias?
and he said I am not. Art thou the prophet? and.
he answered no.*"

" This conversation a writer used to composition
would rather have related in a more connected manner, as follows. *Then the Jews sent Priests and
Levites from Jerusalem to ask him who he was,
and he confessed he was not the Christ: They asked
him if he were Elias, but he said he was not, if he
were that prophet, but he answered no.*"

" Upon

" Upon these principles we may perhaps be able
to give a more complete solution than hath hitherto
been given of a paradox in the history of letters : viz.
Why, generally, the first regular compositions of any
people should be in *verse*, rather than in *prose*.   One
reason, no doubt, was that, antecedent to the use of
letters, verse was much more proper than prose in
compositions that were designed to perpetuate the
memory of remarkable transactions and events, as
deviations from the original would be made with
more difficulty, and corrupted passages restored with
more ease : But, additional to this, we may perhaps
affirm with truth, that the concatenation of sentences
is not so intricate in verse, as in prose.   Not unfre-
quently the neglect of regular transitions is esteemed
graceful in verse and the old poems here referred to,
as the *Delphic Oracles*, *&c.* where the sense was generally compleated in a line, or a short stanza, required very little art or variety of connexion.   How
much more elaborate in point of transition and concatenation of sentences is even the history of *Herodotus*
than the poems of *Homer*, many parts of which are
historical."

Lecture 12th.   *On the growth and corruption of*
Langua-

Languages. All languages whether antient or modern, are subject to growth, improvement, and decline, as well as to many intermediate fluctuations.
The causes of these are extraneous, and no internal
structure of the language can prevent these changes.
They will arrive at their acmè sooner, and be more
perfect and copious in proportion as the people are
more literary, more mercantile and enterprizing : for
such a people having more ideas will require more
words to express them.   Hence the superiority of
the Greek to the Hebrew.   At this stage the decline
of the language usually commences.

   " The progress of human life in general is from
poverty to riches, and from riches to luxury, and
ruin : in *Architecture* structures have always been at
first heavy, and inconvenient, then useful and orna-
mental, and lastly, real propriety and magnificence
have been lost in superfluous decorations.   Our very
dress is at first plain and aukward, then easy and elegant, and lastly downright fantastical.   Stages of a
similar nature may be observed in the progress of all
human arts ; and language, being liable to the same
influences, hath undergone the same changes.
Whenever a language hath emerged from its first
                                                rough

rough state of nature, and hath acquired a sufficient copia of significant and harmonious terms, arbitrary and whimsical ideas of excellence have been super-added to those which were natural and becoming, till at length the latter have been intirely sacrificed to the former."

These observations he illustrates by a short history of the revolutions of the Roman language.

Dr. Priestley seems to think that there is a period when the language of a nation will no longer admit of improvement, viz. when power and influence abroad, and arts, science and liberty at home, have arrived at their greatest height. But when has this been? And who can say when it will ever be? It is to be hoped, never. He thinks one entire century favourable to the polite arts, will suffice to bring any language to its perfection : and that the English was fixed in the reign of Queen Anne, But this has certainly not been the case with the English language, which Mr. Godwin has fully shewn to have been hitherto progressive; contrary to the opinion of most writers before him. *Enquirer.*

The rest of this lecture consists of observations on Academies ; on the Analogy, and on the Standard of Languages.

The

The 13th and 14th are *on the complex structure of the Greek and Latin Languages.* These discourses will not readily admit of analysis, and I do not observe any passage sufficiently striking to introduce it here, excepting the following extract which closes the 14th lecture.

" But the present *Italian, French* and *Spanish* tongues, most probably, took their rise from the imperfect attempts of barbarous nations to speak the Roman Tongue, and particularly in the provinces; and that, long before the dissolution of the Roman empire by the irruption of the northern nations."

" If we consider the Grammar of those languages with attention and compare them with the Latin, we may almost see the very manner in which they were produced. Foreign nations, in attempting to speak Latin, could not avoid imitating the principal tenses of their verbs: accordingly we can plainly discern the form of them in their present languages. When people who had not the advantage of a regular and perfect instruction endeavoured to speak in Latin, they would naturally think of nothing but of rendering the words of their own tongue literally into it; and when nations of the *Teutonic* original had rendered

ed

ed into some sort of Latin, or retained, their own articles, casual prepositions, and auxiliary words
(which, being fundamental in their own language,
would be the last things they would part with, and
indeed what they could have no idea of their being
able to do without) they would find that more inflections were unnecessary. The Roman soldiers,
who formed the colonies, being no great masters of
the language, would make no great difficulty of
leaning to this barbarous manner of speaking it.   It
confirms this conjecture, that the present *Italian*,
French and *Spanish* tongues were originally called
Roman, in opposition to the native languages of those
who spoke them."

"*Greece* being continually open to the inroads of the
Italians, *Germans*, *French*, and other northern *Euro-
peans;* particularly about the time of the *Croisades*,
the *Greek* language admitted a good deal of the idiom
of the northern tongues in the same manner : though,
from the forementioned internal causes, it had lost
a great number of its inflections before ; as was most
observable about the time of the emperor *Justinian*,
and this change had begun so early as the translation
of the seat of the empire from *Rome* to *Constantinople*."

In

In the *modern Greek*, we see almost a literal trans-lation of some of the Teutonic auxiliary verbs into Greek, in ειχα for had, and θελω for will; which of course supplanted a great part of their former variety of tenses; for the modern Greeks say ειχα γραψει I had written, ειχας ϛραψει thou hadst written, &c. θελω ϛραψει I will write, θελεις ϛραψει Thou wilt write, &c: and to supply the place of moods, they have evidently translated their own forms of expressing the modes of affirmation into Greek particles, which they have pre-fixed to the common inflections."

Lecture 15th. *Of the revolutions of Language, and of Translations.* When nations are conquered, it has generally been the case that the conquered na-tions especially if dispersed, loses its language; as was the case with the Jews after the Babylonish irrup-tion and captivity. Thus the English gains ground on the Irish, the Erse, and the Welch; and the French on the Britannoise. If the conquerors mix with the inhabitants of the conquered countries in great num-bers, the languages will be mixed, or new ones formed; as the Italian, Spanish, &c. from the Ro-man. Where the conquered nation had arrived at considerable eminence in arts and literature, the lan-guage

guage of the conquered country will be adopted a-
mong the better informed classes of the conquerors,
as the Greek language prevailed among the literati
at Rome: for the knowledge possessed by the
Greeks, must have been sought after by means of
the language in which it was written and preserved.
On these principles the Latin language seems to be
adopted by the learned of Europe.* The second
part consists of general remarks on the mode and
use of translating.

Lecture 16th. *Of Metrical Compositions.* " The
first verses (like the rudiments of all other arts) were
probably made by chance. The harmony of words,
at first casually placed in metrical order, would engage the attention. The pleasing sensation accom-
panying it would excite mankind, when they were
first at leisure to attend to their language, to consider
the

---

* On these two principles combined, it may be well explained, how
the English language comes to be a mixture of Saxon (the original) of
Danish and Norman French, (forced into practice by the conquerors)
of Latin and a small portion of Greek, voluntarily adopted by the literary class, and gradually incorporated with colloquial forms of expression. T. C.

the nature and manner of it; from whence the transition to imitation is universally natural."

" When verse became tolerably familiar and easy, and before the art of writing was invented, it would soon be perceived to be an excellent vehicle to convey the knowledge of past transactions to posterity; since verses are easily committed to memory, and the regularity of the measure helps to prevent mistakes in the repetition. Thus, in time, would most nations become stocked with traditional poems, serving for memorials of remarkable transactions; of which those relating to their Gods and Heroes would be repeated, and sung in their honour, at their festivals. For the invention of *Musick* and *Poetry* hath, in all nations, been nearly cotemporary; and there have always been methods of adapting the one to the other. The simple pronunciation of the ancients being slow and raised, must of itself have been musical."

" Things being in this situation, it is natural to suppose, that the first things men would think of committing to writing (after the art was invented by them, or communicated to them) would be these *poems;* and it might be some time before they would think of making use of the art for any other purpose.

Accord-

Accordingly, we find in history, that, in most nations, poems were the first compositions that were committed to writing, and that, the art of prose-writing was subsequent to it. Sir *Isaac Newton* (I suppose upon the authority of *Strabo*) says that the *Greeks* wrote nothing in prose before the conquest of *Asia* by *Cyrus*, about which time *Pherecydes Scyrius*, and *Cadmus Milesius* introduced writing in prose."

The following remarks on the metre of the antients, in the application of music to poetry appear to be just.

" All the harmony that the *Antients* ever attempted to give their language, arose from the proper disposition of the syllables according to their *quantity*, as divided into *long* and *short*; two short syllables requiring the time of one long one. To exemplify this, take the following verses of *Virgil.*"

Tityre, tu patulæ recubens sub tegmine fagi
Sylvestrem tenui musam meditaris avena.

" All the harmony of these verses consists in the proper disposition of the long and short syllables. And, according to the more elaborate pronunciation of the ancients, the difference in the length of syllables would be more observable than it is with us. This

regard

regard to quantity did not in the least interfere with the raising or depressing of the voice, which they called *Accent.*"

" On the contrary, according to our method of pronunciation, it is impossible to distinguish the quantity and accent. We pronounce every syllable with equal rapidity, except one syllable in every word, which we pronounce with more force than the rest; which, doth, in some measure, of necessity, occasion a protraction of the sound. It is the regular fall of this accent that constitutes the principal part of the harmony of all *European* verses, as in the following in *English*."

Say why was man so eminently raised
Amid the vast creation, why ordained
Through life and death to dart his piercing eye,
With thoughts beyond the limits of his frame ?

" Besides this another kind of harmony hath been introduced into most modern languages; which is the similarity of sound at the close of the verses, called *Rhyme*. The following have this kind of harmony."

Know then this truth (enough for man to know)
Virtue alone is happiness below.

The

The only point where human bliss stands still,
And tastes the good without the fall to ill :
Where only merit constant pay receives,
Is blest in what it takes and what it gives.

POPE.

" The principle therefore, or source of harmony, in ancient and modern poetry, is totally different : the former arises intirely from quantity, the latter from the accent; and sometimes accent in conjunction with rhyme."

" For this reason the ancient poetry was, of the two, the better adapted to musick ; which is regulated chiefly by time ; and perhaps the just pronunciation of verses, according to the natural length of the syllables (with a peculiar raised tone of voice, and such variations with regard to acute and grave as that manner of speaking would naturally throw the voice into) might be that in which the principal part of ancient vocal music consisted. Whereas, in modern music, (unless a long note be contrived to receive the accent through the whole verse, which is seldom done, except in some few songs) our poetry hath no more than an arbitrary connection with our music, and prose suits it quite as well : but to sing prose

B b 3

would

would have been reckoned very absurd among the *Ancients*, it being a thing that was never thought of or attempted by them."

" Our music, indeed, may be contrived to correspond, in general, to the sentiment and passion expressed in a poem : for as the verses may be of a diverting or mournful nature, the music may likewise, upon the whole, tend to inspire mirth or melancholy ; but the particular words of the poem have still no real connection with the particular strains of the music. How often do we see, in very good musical compositions, where words are subjoined, the most expressive and important strains in the music to fall upon very trifling words in the poem. And do we not generally sing the same notes to every stanza of an ode ; notwithstanding that the variation of the sentiment, and the different disposition of the emphatical words in the line, seem to require a proportionable change in the notes that are sung with it : yet so arbitrary and general is the connexion between our music and our poetry, that the absurdity is not perceived."

" Modern languages, and *English* in particular, do not admit of the measures of ancient poetry ; because

cause the distinction of our syllables into long and short is not sufficiently apparent. According to the rules of ancient versification, too great a number of them would be long. On the other hand, the least tendency to rhyme was condemned as vicious in ancient poetry ; till, in some late centuries, when the ancient pronunciation of the Latin was forgotten, some *monks* composed Latin verses in rhyme, but without any regard to the quantity. One of them is said to have turned the whole *Æneid of Virgil* into rhyme."

" As the rules of versification do necessarily confine a writer in the choice of his words, poets, in all languages, take liberties which are not allowed to prose writers. This is called the *licentia poetica*, and makes the language of verse differ considerably from that of prose. In the *Italian* tongue this is very observable : for instance, *anima*, in that language *the soul*, in prose ; when, in verse, it is changed into *alma*."

There have been attempts at introducing the Sapphic measure into the English language, by Watts, and Southey; and Collins's metrical ode to the evening has found some imitators, and I think I re-

collect

collect some arythmic hexameters, but rhyme combined with metre seems most natural to the language. Whether the ingenious author of " Metronariston " has not far over-rated the pleasure to be obtained from his method of reading the Greek and Latin poets, can hardly be judged of, but by submitting the experiment to the ear. How the antients pronounced their words or even their letters we cannot now tell, and therefore I shall not be able to join in opinion with Dr. Priestley in a passage to be quoted presently, that the Greeks *did not* pronounce certain characters as the moderns do. Dr. Parnell's imitation of a part of Pope's Rape of the Lock may serve as a favourable specimen of the effect of monkish rhyme (rhyme intermixed with Hexameter measure) on the ear; and the anacreontic of Walter de Mapes, preserved in Cambden's Remains, and partly translated by Huddisford, affords a tolerable example of another variety of Latin rhyme; but without affording at the same time, any temptation whatever to pursue the practice.

The 17th Lecture consists of *Observations on the different properties of Language.* The perfection of a Language, consists principally in having a sufficient

ficient copia of words, in the absence of ambiguity, and in a pleasing, not harsh or grating pronunciation.

On these three points of excellence he enlarges. The copiousness of a Language, depends *chiefly* on the state of improvement among the people that use it. But this is relative, for some people may be improved in one, and another in other respects. Thus people like the Greeks who cultivate poetry and oratory to a high degree, are likely to have their language abound in synonimous and nearly synonimous terms ; whereas a nation cultivating the arts and sciences extensively, will of course require terms appropriate to new objects and combinations. Thus the number of words in the English Language is not greater than in the Greek.

Ambiguity may arise from the same word having more ideas annexed to it than one. And from want of means to ascertain the relations of words to each other.

The harshness of a Language, does not absolutely depend on the mere proportion of consonants to vowels, but more on their arrangements : thus the words *strand*, *blind*, do not sound harsh at least to an English ear. Neither do the words, *that*, *then*,

thread,

thread, &c. although our neighbours complain so much of the *th*; were they to learn the use of it, the complaint would cease. " The Hebrew, Ara-" bic, and other Eastern Tongues, are thought very " sweet and melodious in the countries where they " are or have been spoken, and yet they abound with " *gutturals* and *aspirates*, which are in their own na-" ture, the most difficult and violent articulations " that the human voice is capable of. In the Greek, " frequent use is made of χ (answering to ‏ח‎ in He-" brew); also of ϑ and φ (none of which were pro-" nounced by the Antients as we now pronounce " them) and yet all people think the Greek to have " been a very harmonious musical Language."

It is certainly within the compass of possibility, that the Greeks *did not* pronounce these characters as we do; but I should be glad to know how the Doctor became acquainted with this fact? or how it can be ascertained? I have not here the opportunity of consulting Mekerchus, or Beza, or Vossius,* or

Lipsius,

---

* *Adolphus Mekerchus*, de veteri et recta pronuntiatione Linguæ Græcæ. *Vossius* de poematum Cantu. *Beza* de veteri et Germana pronuntiatione Linguæ Græcæ. *Lipsius* de recta pronuntiatione linguæ Latinæ; dedicated to Sir P. Sydney.

Lipsius, or the Accentus redivivi, or any of the numerous Greek Philologists, but I see not how any man can be certain of this, who did not live in the days of the antients as well as in those of the moderns : especially as the Greeks and Latins offer no Rhyme to guide our conjectures. I remember a conversation between Dr. Johnson and Mr. Dagge during an interval in the performance of Horace's carmen seculare, when set to music by Philidor, and performed under his and Baretti's direction. The subject was, the proper method of pronouncing the Latin language. Johnson. " Sir, this is a question that cannot be settled in this day; no modern can have heard the antients speak ; therefore no modern can tell how the antients spake. One man may instruct another in proper diction by example, but the instruction must be addressed to the ear, not to the understanding; written precept is insufficient. All we can do in the present case is to conjecture, and of conjectures we are bound by the most probable. That the pronunciation of the Latin would be modified and altered by the intermixture of barbarians who overturned the Roman empire is certain ; but in what instances and to what degree is uncertain.

tain. It is probable that the immediate descendants of the Romans would be more likely to pronounce the Roman language with propriety, than
foreign nations. It is probable that persons living in
the same climate, and on the same spot, would be
more apt to fall into the pronunciation which a Roman would adopt, than any foreigner; for the natural causes that affect pronunciation, would be common to the antient and the modern inhabitant of the
same place. For these reasons, I incline to think
that the Italians have the chance of being more correct than any other nation. Another observation
occurs to me, which though it will not decide the
question, will serve to illustrate the arguments I
have employed. When Virgil describes the Cy-
clops as forging the arms of Æneas, he uses language evidently meant to convey a correspondence
of the sound to the sense.

Illi inter sese, magnà vi, brachia tollunt,

Innumerum : versantque tenaci forcipe ferrum.

Pronounce this passage like an Englishman, and
the beauty almost vanishes : pronounce it like an
Italian, and it must be felt."

I think with Johnson, that descendency and similarity

larity of climate, though not conclusive evidences in favour of right pronunciation, as we know they are not either with respect to the modern Romans, or modern Greeks, are yet much stronger than any other people can adduce : and where one mode of pronunciation is universally adopted, it has a higher degree of probability in its favour than any other can pretend to.

Dr. Priestley proceeds to remark that whether a language is harsh or not, must be judged of from the *best* writers in it : for there may be more difference between two writers in this respect than between two languages.

Also, that the real structure of an harmonious language must admit of any words or numbers of words to succeed each other with ease as if they were one word. Hence there must not be too many consonants thrown together at the beginnings and endings of words : else they will impede facility of pronunciation. Having made these preliminary observations, he proceeds in the 18th Lecture to a *Comparison of different Languages.*

In this Lecture he briefly considers the characteristic differences of the Hebrew and the Greek languages,

guages, adding some short remarks on the Latin, French, Italian, Spanish and German languages. The first part in particular is interesting, brief as it is : but the plan of this account will hardly author-ize the transcribing of it here.

The 19th Lecture is on the *origin, use, and cessati-on of the Diversity of Languages.*

The present diversity of languages is the necessa-ry effect of the new wants and new situations in which mankind would gradually find themselves. There must have been a first or original language : this the Scriptures teach. But that language consisting of few words, and of few inflections because few would be needed, could easily be altered so much as to become a different thing from what it was original-ly. This is far more probable than any miraculous interference at the building of Babel. The difficul-ty of conceiving how languages should be so numer-ous and so different, rests upon the supposition that the primitive language was copious and perfect; but suppose it no more so than was necessary to primitive wants, the difficulty no longer remains.

Observations succeed on the utility of different languages, and the necessity of attending to the construction

struction of more than one, by those who wish accurately to understand their own.

On a Philosophical Language: Sketch of Dr. Wallis's* plan. Doubts whether a distribution of of things and characters into classes, can be managed sufficiently well, in the present state of knowledge : whenever the present diversity of languages has sufficiently answered all the beneficial purposes for which it was ordained or permitted : whether the theory of languages itself as an abstract science, be sufficiently advanced, to enable us to frame a philosophical language and character, that will answer the proposed ends. But he thinks, that when the present diversity has continued so long as to be *functa officio*, it will gradually bring in the necessity of such a language as has been proposed.

The volume closes with a list of the books he made use of, viz. the Grammars of Messieurs de Port Royal. Harris's Hermes. Bayley's Introduction to Language. Robertsons method of reading Hebrew.

This is a misprint for *Wilkins*. Dr. Wallis's Grammar and his Dissertation de Loquela seu Sonorum formatione is curious, and appears to have been practically applicable to the teaching of deaf persons to speak.

Hebrew. Hartley on man. Du Fresnès Glossary of modern Greek. Reland's Miscellaneous Dissert. Richards's Welch Grammar and Dictionary. Wilkins's Essay toward a real Character and Philosophical Language. Brerewood on Language, and Sharpe's two Dissertations on Language.

Had he revised these Lectures, with the advantage (in addition to much more reading and reflection) of Mr. Horne Tooke's labours, and the books referred to by him, and some few others easily obtained, they would have been well worth the attention of the public in such an improved state: as they are, I know of no treatise so well adapted to the purposes for which it was composed and compiled.

I have been the more diffuse on this work of Dr. Priestley because it does not appear to have been much known beyond the circle of his students. The printed copy he kept by him, has spaces left for the Greek and Hebrew quotations which he has inserted in his own hand writing, with a few corrections and additions in short hand. Indeed he has mentioned in the prefixed advertisement, that if these Lectures should happen to fall into other hands than those for whom

they

they were intended, they must only be regarded as the heads of discourses to be enlarged upon by the Lecturer at the time of delivery.

The Lectures on Oratory and Criticism, and the Lectures on General History and Civil Policy, of which last a new edition has lately been published in two volumes, at Philadelphia, are too well known to require to be enlarged upon. In the former the only peculiarity seems to be the adoption of Hartley's Theory of Association to explain and illustrate many of the subjects treated, to which no doubt that theory is well fitted.

The Lectures on History and Civil Policy have been so well received by the public, and they treat of subjects so important, and contain such a mass of information, that they must long remain a stock-book to the student.

In the last edition of these Lectures, the Dr. has inserted a new chapter on the constitution of the United States (chap. 43) in which among other observations he has introduced the following.

" To this view of the constitution of the United States I shall take the liberty to subjoin a hint of what appears to me to be of particular importance as a

C e   *maxim*

maxim of policy in the present state of the country in general, though I have enlarged upon it on another occasion, it is not to favour one class of the citizens more than another by any measure of government, especially the merchant more than the farmer."

" Their employments are equally useful to the country, and therefore they are equally entitled to attention and protection, but not one more than the other."

" If the merchant will risk his property at sea, let him calculate that risk, and abide by the consequence of it, as the husbandman must do with respect to the seed that he commits to the earth; and let not the country consider itself as under any obligation to indemnify one for his risks and losses any more than the other, especially as, in the case of the merchant, it might be the cause of a war with foreign states.   If there should be danger from the depredations of privateers, or ships of war of any other kind, let the merchants have the power of defending their property, and let them and the insurers indemnify themselves, as they always will do, by the advanced price of their goods, but in no other way whatever.   If in defending themselves they offend other nations, let them be

given

given up to punishment as pirates. If the risk of a national quarrel be manifest, let the trade be prohibited."

" If the expence of fitting out fleets for the protection of any branch of commerce exceeds the advantages that arise to the country from that commerce; there cannot be any wisdom in prosecuting it. In that case let that branch of commerce be abandoned; and it may be depended upon that the country will not long be in want of any valuable commodity with which the merchants of other countries can supply it, and that the competition will prevent the price from becoming exorbitant."

" No proper *merchandise*, or the peculiar advantage of it, would be lost by this means; but only that particular branch of industry and gain called the *carrying trade*, which would be left to other nations that could carry it on to more advantage; while the exchange of commodities, that of the articles that the country can spare, for those that it wants, would be the same as before; and the capital that had been employed in the carrying trade might be employed to more advantage some other way, of which the holders will be the best judges."

This

This important subject, he has treated more at length in a paper published in the " Aurora " signed a Quaker in politics, which is subjoined to this Appendix.

This subject was afterwards treated somewhat more systematically by Mr. Cooper in a paper published among his Essays, and the same general ideas have been advanced by Arthur Young, Esquire. It would be well for the rising generation of this continent, if the momentous question so discussed, were as fully considered by our legislators and statesmen as its importance deserves.

I believe the Chart of Biography, is an invention to which Dr. Priestley has the sole claim, and a beautiful specimen it is, of the aid which memory may derive from mechanical contrivance. Dr. Gray's *Memoria technica*, though ingenious, is still a great exertion on the memory, from the number of harsh and arbitrary sounds of which the verses according to his plan must necessarily be composed : and the missing or mistake of a single letter is fatal to the required accuracy. In this map, a glance of the eye takes in not only the period of life of the person who is the object of enquiry, but that of all his cotemporaries.

raries. This chart has had tolerable success : a new and improved edition of it has been engraven in this country : but it is not yet so general an article of furniture in a literary room, as it deserves to be.

The Chart of History, is an improvement by Dr. Priestley on a French plan of the same kind, and is doubtless of great use as exhibiting at one view a number of the most important general facts of history in connexion with each other, and as suggesting many reflections which would not so obviously occur on the perusal of history in detail. It is also very convenient as an historical compend for occasional reference.

The general idea of these charts has been since adopted and applied with great ingenuity by Mr. Playfair to the rise and progress of national debt, and I believe of national import and export. It might be extended to many other objects of statistic importance, and suggest reflections at a small expence of labour, which might never arise in any other way.

One of the last papers written by Dr. Priestley and which seems to belong to the miscellaneous class of his writings was a letter to Dr. Wistar in reply to Dr. Darwin's observations on Spontaneous Vitality.

Dr.

Dr. Darwin had made use of Dr. Priestley's experiments on the *confervæ fontinalis*, a green matter produced on stagnant water, as favourable to the hypothesis of equivocal generation ; whereas Dr. Priestley who was always of opinion that if a mite could be thus produced, so also might a mammoth or a man, deemed this revival of an exploded hypothesis a direct introduction to Atheism.

Certain it is, that if we argue from facts that we do know, to similar facts with which we are not so well acquainted, a mode of philosophizing undoubtedly legitimate, the preponderance of probability is against the notion revived by Dr. Darwin.* Still however many facts concerning the generation of the smaller animals as insects and animalcules are so perfectly anomalous, as in the case of the Aphis and we know so little on this subject as it respects this whole class of organized beings, that our analogies drawn from

the

---

* I do not recollect any late author of credit who has leaned to the doctrine of equivocal generation beside Darwin, except Mr. Bayley in his Morbid An. tomy : nor do I wonder that Mushrooms, Hydortids, and all the tribe of worms that generate in the viscera of the larger animals should suggest, for a while, some doubts of the more modern and popular hypothesis.

the production of the more perfect when applied to
the less perfect orders of animal life, as the earth
worm, the polypi, the nydra, the millepes and the
whole class of Zoophytes, that room may still be left
for reasonable doubt. Rousseau very properly observes that a philosopher has frequent occasion to
say *J'ignore* but very rarely *c'est impossible*. I do
not see the *certain* tendency of this opinion to atheism, for this property of spontaneous production may
have been originally communicated under certain circumstances as well as any of the other properties of organized or unorganized matter; and the one and the
other may be equally necessary parts of the pre-esta-
blished order of things. But if it do lead to Atheism, what then? There can be no crime in following truth wherever it lead, and I think we have sufficient reason upon the whole to believe, that the result
of truth must be more beneficial to mankind than
error. Nor can I see how the belief of no God can
be more detrimental to society or render a man less
fit as a citizen than the belief of the thirty thousand
Gods of the Pagans,* or the equal absurdities of tri-

---

* I believe the learned mystic and pagan of modern days, Mr.
Taylor is in moral deportment a pattern to his Christian compeers.

nitarian orthodoxy.   It is very dubious whether the
practice (the profane practice I might safely say) of
resorting

Who would not prefer the dispositions of this man, as far as they are
known, to the sneering, sarcastic, the insolent and the intolerant
Bishop of Rochester? I cannot suspect this Hierarchist of having
perused either the Phœdo of Plato in the original or the commentary
of Olympiodorus, especially since his parade of Zuicker, whose
works it is highly probable he had never seen.   But I cannot help
suspecting he had seen the following quotation from the commentator
above mentioned, which I produce for the amusement of the reader
as proper companion to the Bishop's notion of the origin of Jesus
Christ, the second Person in the Trinity.   In his charge to the Clergy of St. Albans, the then Archdeacon (a Saint in crape; but twice
a Saint in Lawn!) says, p. 55.   " The sense of Athenagoras is, that
" the personal existence of a divine logos is implied in the very idea
" of a God.   And the argument rests on a principal which was com-
" mon to all the platonic fathers, and seemed to be founded in Scrip-
" ture, *that the existence of the Son flows necessarily from the father's*
" *contemplation of his own perfections.*   But as the Father ever was,
" his perfections have ever been, and his intellect has been ever ac-
" tive.   But perfections which have ever been, the ever active intel-
" lect must ever have contemplated, and the contemplation which
" has ever been must ever have been accompanied with its just effect,
" the personal existence of the Son."

    Admirable logician! how clearly does this explanation unfold all
the mysterious process of God the Father begetting God the Son,
who it is to be presumed in some similar fit of contemplation begat
                                                                    God

resorting on all occasions to oaths, has done more good than harm: and if society cannot offer within itself sufficient sanctions of reward and punishment,

by

God the Holy Ghost! What a pity these platonic Trinitarians should stop so soon? for the same means would doubtless have furnished us with deities in abundance. The Pagans had 30,000 Gods, why should the Christians content themselves with three?

This passage I long deemed unique, until I perused the DISSER-TATION ON THE ELEUSINIAN AND BACCHIC MYSTERIES, which can acknowledge no author but Mr. Taylor. From many pages equally instructive. I copy the following translation of Olympiodo-rus, who *beautifully observes*, (says Mr. Taylor,) "That these four " governments obscurely signify the different gradations of virtues; " according to which our soul contains the symbols of all the virtues " both theoretical and cathartical, political and ethical; for it either " energizes according to the theoretic virtues, the paradigm of " which is the Government of Heaven, that we may begin from on " high; and on this account Heaven receives its denomination " $\varpi\alpha\rho\alpha$]ov]α $\alpha\nu\omega$ $o\rho\alpha\nu$ from beholding the things above; or " it lives cathartically, the exemplar of which is the Saturnian " Kingdom: and on this account Saturn is denominated, *from being* " *a pure intellect through a survey of himself*; and hence he is said to " devour his own offspring, signifying the conversion of himself to " himself: or it energizes according to the politic virtues, the sym- " bol of which is the government of Jupiter, and hence Jupiter is the " Demiurgus, so called from operating about second natures, " &c. &c."

I be

by positive institutions, and the still stronger influence of public sentiment and common interest, very little can be added in reality, by a recourse to the terms of eternal damnation. However well founded and useful the doctrine of a future state may be, it certainly was no part of the sanction proposed by the antient legislators. Dr. Sykes in his examination of Warburton's paradoxes shews this, with respect to the laws of Zaleucus, Charondas and Cicero, as well as those of Triptolemus, Draco, Solon, Lycurgus and Numa.

The modern and more acccrate notions respecting matter, imperfect as they yet are and probably ever will be, favour the opinions of Dr. Darwin much more than the old doctrines on this subject. Matter is no longer treated of as the sluggish inert substance it was heretofore considered. Whatever be the substratum of its properties, we know those properties

to

I believe I may have alluded elsewhere to this theory of divine generation propounded by the Reverend Bishop, but the subject of the present section forced this precious pair of parallel passages irresistibly on my fancy. I must plead with the poet.

To laugh, were want of Goodness and of Grace,
But to be grave, exceeds all powers of face.

to be highly, perpetually, and essentially active: entering continually into new combinations, and never for a moment permitting any aggregation organized or unorganized to continue precisely the same as at any given previous portion of time. Nor can the probability be denied, that there is a *nisus* to improvement in all organized beings, at least where that organization is attended with the slightest portion of volition, and the power of voluntary effort.

By these observations however I by no means wish to be understood as defending the doctrine of Equivocal Generation, which Dr. Darwin's ingenuity has again introduced to the notice of the philosophical world. But though the balance of probability may be on the side of the more fashionable opinion, I cannot help thinking that a candid observer may even yet be allowed to doubt.

Dr. Darwin seems to think that Dr. Priestley's green matter could not arise from seeds diffused through the air but must be generated in the water wherein it is observed. To ascertain this, Dr. Priestley, (who believed that all the parts of the plant or the animal pre-exist in the embryo and are merely extended, not formed anew by
nutri-

nutrition) on the 1st of July 1803 placed in the open air several jars of pump water, two of them covered with olive oil, one in a phial with a ground glass stopper, one with a loose tin cover, and the rest with the surface of the water exposed to the atmosphere; and having found that the addition of vegetable matter aided the production of the *conferva*, he put twenty grains of sliced potatoe into each of the large vessels, and ten grains into each of the smaller. Into each of two very large decanters, the mouths of which were narrow, he put fifty grains: one of these had oil on the surface and the other none. He also filled a large phial with the same water, and inverted it in a vessel of mercury. In about a week the wide mouthed open vessel began to have green matter and the large decanter with a narrow mouth had the same appearance in three weeks. On the 1st of August the vessel which had a loose tin cover extending about half an inch below its edge, began to shew a slight tinge of green; and on the 1st of September the phial with a ground stopper (but which did not fit exactly) began to have green matter; but none of the vessels that were covered with oil, or had the mouth inverted in mercury shewed any such appearance.

ance. On the 11th of September having waited as he thought long enough, he put an end to the experiment.

Here then the access of air was evidently necessary to the production of this green matter, and in the stopt decanter, the seed must have insinuated itself through a very small interstice, and in the decanter covered with a tin cover, it must have ascended and then descended into the water. These facts Dr. Priestley regards as hostile to the hypothesis adopted by Dr. Darwin.

For the other observations on this subject which Dr. Priestley makes in his letter to Dr. Wistar, the reader must be referred to the 6th volume of the American Transactions, wherein it is, or will be printed.

I confess *(pace tanti viri)* that these experiments do not appear to me to be conclusive. The access of air seems almost universally necessary as a stimulus to animal life in the cases which we are well acquainted with, though some of the insect tribe seem to furnish exceptions. The oxygenated arterial blood is evidently conveyed to the infant by the umbilical vessels and placenta: nor do we know decidedly

edly of any animal or plant that can live without access of air. No wonder therefore that the same
might be the case with the plant in question. The
subject deserves more consideration by means of direct experiment than has yet been given it.

As to the opinion to which Dr. Priestley seems to
incline in common with Haller and Bonnet and Spalan-
zani, that the original seed contained the embryons
of all future plants, and that our first mother Eve bore
in her ovaria every individual of the human race, like
a nest of boxes in a turner's shop, one within the o-
ther (Emboitement as Bonnet calls it) I cannot think
it will maintain its ground. To suppose that five
or six hundred thousand millions of human creatures were thus pent up all perfect and ready formed,
in the small compass assigned for their reception in
the first female parent, is so pregnant with absurdity,
that the relations of Bishop Pontoppidan are as the
axioms of Eucli to it. I have not seen Blumenbach's
work on generation, nor do I know whether the con-
ferva fontinalis on which he experimented, was the
green vegetable matter of Dr. Priestley. I agree
however to the ridicule which he throws on this system in the extract which Dr. Willich has given

(Lect.

(Lect. p. 376 ed. Boston) and I think his plastic *ni-sus* is sufficiently near to the spontaneous vitality of Dr. Darwin to class these philosophers together so far as the present subject is concerned.

The Following.

FUGITIVE PIECES BY Dr. PRIESTLEY,

Are deemed sufficiently interesting to be preserved; and as two of them have hitherto been published only in a Newspaper or a Magazine, they are inserted here as properly belonging to the class of his Miscellaneous works. The paper concerning Mr. Burke, was prepared by Dr. Priestley for the Press but a few days before his death, and has not hitherto appeared in print.

MAXIMS OF POLITICAL ARITHMETIC,

Applied to the case of the United States of America, first published in the Aurora; February 26 and 27, 1798. (By a Quaker in Politics.)

AN idea of the true interests of any country is perhaps most easily formed by supposing it to be the property of one person, who would naturally wish to derive the greatest advantage from it, and who would therefore, lay out his capital in such a manner as to make it the most productive to him. An attention to the separate and discordant interests of different clas-

ses

ses of men, is apt to distract the mind : but when all the people are considered as members of one family, who can be disposed of, and employed, as the head of it shall direct, for the common benefit, that cause of embarrassment is removed.

To derive the greatest advantage from any country it will be necessary that attention be paid, in the first place, to the wants of nature, and to raise from it, in the greatest quantity and perfection, such producti- ons as are necessary to feed and clothe the inhabitants, and to provide them with habitations, in order to guard them against the inclemency of the weather ; and after this such as are of use to their more com- fortable accommodation, and the supply of artificial wants.

If any country be completely insulated, or cut off from all communication with other countries, it will be necessary to raise all those articles within itself; but when a communication is opened with other countries, the proprietor will do well to give his whole attention to those productions which his own coun- try can best yield, and exchange the surplus for such articles as other countries can better supply him with. For by that means, his labour will be employed to the

D d   most

most advantage. If, for example, it would employ him a month to go through all the processes which are necessary to make a piece of cloth, when the effect of the labour of a week in his husbandry would enable him to purchase that cloth, it will be better for him to confine himself to his husbandry, and buy his cloth ; besides that, not making it his sole business, he would not, with any labour, make it so well. And now that a communication by sea with all parts of the world is so well established, that it may be depended upon that whatever any country wants another can supply it with, to the advantage of both, this exchange may be made with little interruption, even by war.

Commerce consists in the exchange of the commodities of one country for those of another ; and as this, like any other business, will be performed to the most advantage by persons who give their whole attention to it, and who are called *merchants*, it will be most convenient, in general, that this be done by them, rather than by those who employ themselves in raising the produce. The business of conveying the produce of one country to another is a different thing from merchandise. Those who employ ships for

for this purpose, are paid for their trouble by the freight of their vessels, while the merchant subsists from what he gains by the exchange of commodities.

What is generally termed *active commerce* is that which is carried on by the natives of any country in ships of their own, conveying their produce to other countries, and bringing back theirs in return ; and that is called *passive commerce* which is carried on at home, people of other countries bringing their commodities, and taking back what they want in exchange for them. The quantity of proper commerce, or merchandize, is the same in both these cases. All the difference consists in the employment given to the carriers, and the shipping, of the different countries.

While the communication with other countries by sea is open, it cannot be for the interest of any country, either to impose duties on goods brought into it, or to give bounties on those that are exported : because, by both these means, the people are made to pay more than they otherwise would do for the same benefit. In both cases the price of the goods must be advanced. He who pays the duty will be refund-

ed

ed at least, by the persons who purchase the commodity, and the bounty to the vender must be paid by
a tax on all the inhabitants.

It is, no doubt, the interest of any particular class
of persons to extend their business, and thereby increase their gains. But if their fellow citizens pay
more in the advanced price of what they purchase
than their gain amounts to, the community is a loser;
and if it be equal, one class is made to contribute to
the maintenance of another, when all have an equal
natural right to the fruits of their own labour.

For the same reason, if, on any account, the conveyance of goods from one country to another be attended with more loss than gain, the person in whose
hands was the property of the whole would discontinue that branch of business, and employ his capital
in some other way, or rather let it remain unproduc-
tive than employ it to a certain loss.

These maxims appear to me to be incontrovertible
in the abstract. What then may be learned from
them with respect to this country, situated as it now
is?

Without enquiring into the cause, which is no part
of my object, it is a fact, that the conveyance of
goods,

goods, or the carrying trade of this country, which
has generally been taken up by the merchants, though
it is no necessary branch of their business, is peculiarly hazardous, and of course expensive. This expence the country at large must pay, in the advanced
price of the goods purchased. In this state of things
it has also been found necessary to send ambassadors to distant countries, in order to remove the supposed cause of the difficulty, which is attended with
another expence. It has likewise been thought necessary to build ships of war for the purpose of pro-
tecting this carrying trade; and if this be done to
any effect it must be attended with much more ex
pence.

I do not pretend to be able to calculate the expence
occasioned by any of these circumstances; but the
amount of all the three, viz. the additional price to
the carrier to indemnify him for his risque, the expence of ambassadors, and that of fitting out ships of
war, I cannot help thinking must be much more
than all the profit that can be derived from the carrying trade; and if so, a person who had the absolute
command of all the shipping, and all the capital, of
the country, would see it to be his interest to lay up

his ships for the present, and make some other use of his capital. And as the greatest part of the country is as yet uncleared, and there is a great want of roads, bridges and canals, the use of which would sufficiently repay him for any sums laid out upon them, and they would not fail to contribute to the improvement of the country, which I suppose to be his estate, he would naturally lay out his superfluous capital on these great objects. The expence of building one man of war would suffice to make a bridge over a river of considerable extent, and (which ought to be a serious consideration) the morals of labourers are much better preserved than those of seamen; and especially those of soldiers.

Another great advantage attending this conduct is, that the country would be in no danger of quarrelling with any of its neighbours, and thereby the hazard of war, which is necessarily attended with incalculable evils, physical and moral, would be avoided. To make this case easier to myself, I would consider injuries done by other nations, in the same light as losses by hurricanes or earthquakes, and without indulging any resentment, I would repair the damage as well as I could. I would not be angry
where

where anger could answer no good end.   If one nation affront another, the people would do best to take it patiently, and content themselves with making remonstrances.   There is the truest dignity in this conduct ;   and unprovoked injuries would not often be repeated, at the injurious nation would soon find that it gained neither credit not advantage by such behaviour.

This is the case with independent individuals, and why should it be otherwise with independent nations ? Rash and hasty men, standing on what they fancy to be *honour*, are ever quarrelling, and doing themselves, as well as others, infinitely more mischief than could possibly arise from behaving with christian meekness and forbearance.   In fact, they act like children, who have no command of their passions, and not like men, governed by reason.   In this calculation, peace of mind, which is preserved by the meek, and lost by the quarrelsome, is a very important article.

It will be said, that merchants, having no other occupation than that of sending goods to foreign conntries, by which their own is benefited, have a *right* to the protection of their country.   But what is the *rule of right* in this, or any other case ?   It must

D d 4   be

be regulated by a regard to the good of the whole; and if the country receive more injury than benefit by any branch of business, it ought to be discontinued; and those who engage in any business, should lay their account with the risque to which it is exposed, as much as the farmer with the risque of bad seasons, for which his country makes him no indemnification, though his employment is as beneficial to it as that of the merchant.

If, therefore, in these circumstances of extraordinary hazard, any person will send his goods to sea, it should be at his own risque : and the country, which receives more injury than advantage from it, and whose peace is endangered by it, should not indemnify him for any loss. Let him, however, be fully apprised of this; and if he will persist in doing as he has done, the consequence is to himself, and his country is not implicated in it.

This is a country which wants nothing but *peace*, and an attention to its natural advantages, to make it most flourishing and respectable : and wanting the manufactures of other countries, its friendship will be courted by them all, on account of the advantage they will derive from an intercourse with it. Other

countries

countries being fully peopled, the inhabitants *must* apply to manufactures; and where can they find such a
market as this must necessarily be? And on account
of the rivalship and competition which there will be
among them, the people of this country cannot fail to
be served in the cheapest manner by them all·   This
will be independent of all their politics, with which
this country has nothing to do.   But if, by endeavouring to rival any of them in naval power (which
will only resemble the frog in the fable endeavouring
to swell itself to the size of the ox) it excites their
jealousy, and this country should join any one of
them against any other, it will certainly not only lose
the advantage it might derive from the trade of that
country, but pay dearly for its folly, by the evils of a
state of warfare.

What seems to be more particularly impolitic in
this country, as ill suiting the state of it, is the duty
on the importation of *books*, which are so much wanted, and which even great encouragement could not
produce here.   Is it at all probable that such works
as the Greek and Latin classics, those of the christian
Fathers, the Polyglott Bible, the Philosophical
Transactions, or the members of the Academy of
Sciences,

Sciences, &c. &c. will, in the time of our great grand children be printed in the United States ? and yet there is a heavy duty on their importation ; and for every printer, or maker of paper for printing, there are, no doubt, several thousand purchasers of books, all of whom are taxed for their advantage. In these circumstances, it were surely better to have more cultivators of the ground, and fewer printers.

When I see at what expence ambassadors are sent to foreign and distant countries, with which this country has little or no intercourse ; and when it is very problematical whether in any case, they have not done more harm than good, and think what solid advantage might be derived from half the expence in sending out men of science for the purpose of purchasing works of literature and philosophical instruments, of which all the universities and colleges of this country are most disgracefully destitute ; and that the expence of one of the three frigates would have supplied all of them with telescopes equal to that of Dr. Herschell, and other philosophical instruments in the same great style, to the immortal honour of any administration, I lament that the progress of national wisdom should be so slow,

and

and that our country profits so little by the experience and the folly of others.   The Chinese never
had resident ambassadors in any country, and what
country has flourished more than China?

A foreigner travelling in the interior part of this
country, and finding the want of roads, bridges and
inns, wonders that things of such manifest utility
should not have more attention paid to them, when he
sees that great sums are raised and expended on objects, the use of which is at best very doubtful.   And
men of letters coming to reside here, find their hands
tied up.   Books of literature are not to be had, and
philosophical instruments can neither be made nor
purchased.   Every thing of the kind must be had
from Europe, and pay a duty on importation.

But all this may be short sighted speculation; and
it may be, nay I doubt not it is, better for the world
at large, that this progress should not be so rapid;
that a long state of infancy, childhood and folly,
should precede that of manhood and true wisdom;
and that vices, which will spring up in all countries,
are better checked by the calamities of war than by
reason and philosophy.

It may be the wise plan of Providence, by means
of

of the folly of man, to involve this country in the vortex of European politics, and the misery of Eupean wars; and to prevent the importation of the means of knowledge till a better use would be made of them. Nations make flower advances in wisdom than individual men, in some proportion to their longer duration. But what they acquire at a greater expence, they retain better; so that, I doubt not, there is much wisdom in this part of the general constitution of things.

A stranger is apt to wonder that political animosity should have got to so great a height in this country, when all were so lately united in their contest with a common enemy; and that their enmity, which cannot be of long standing, should be as inveterate as in the oldest countries, where parties have subsisted time immemorial. But it may be the design of Providence, by this means, to divide this widely extended country into smaller States, which shall be at war with each other, that by their common sufferings their common vices may be corrected, and thus lay a foundation for the solid acquisition of wisdom; which will be more valued in consequence of having been more dearly bought in some future age.

Divided

Divided as the people of this country are, some in favour of France, and others of England, I should not much wonder, if the decision of the government in favour of either of them should be the cause of a civil war. But even this, the most calamitous of all events, would promote a greater agitation of men's minds, and be a more effectual check to the progress of luxury, vice, and folly, than any other mode of discipline, and at the same time that it will evince the folly of man, may display the wisdom of Him who *ruleth in the kingdoms of men*, and who appoints for all nations such governments, and such governors, as their state, and that of other countries connected with them, really requires. Pharaoh occupied as important a station in the plan of Divine Providence, as king David, though called *a man after God's own heart*. For his wise and excellent purposes, the one was as necessary as the other.

Many lives, no doubt, will be lost in war, civil or foreign; but men must die; and if the destruction of one generation be the means of producing another which shall be wiser and better, the good will exceed the evil, great as it may be, and greatly to be deplored, as all evils ought to be.

A stran-

A stranger naturally expects to find a greater simplicity of manners, and more virtue, in this *new country*, as it is called, than in the old ones. But a nearer acquaintance with it, will convince him, that, considering how easily subsistence is procured here, and consequently how few incitements there are to the vices of the lower classes especially, there is less virtue as well as less knowledge, than in most of the countries of Europe. In many parts of the United States there is also less religion, at least of a rational and useful kind. And where there is no sense of religion, no fear of God, or respect to a future state, there will be no good morals that can be depended upon. Laws may restrain the excesses of vice, but they cannot impart the principles of virtue.

Infidelity has made great progress in France, through all the continent of Europe, and also in England; but I much question whether it be not as great in America ; and the want of information in the people at large, makes thousands of them the dupes of such shallow writings as those of Mr. Paine, and the French unbelievers, several of which are translated and published here ; and either

through

through want of knowledge, or of zeal, little or nothing is done by the friends of Revelation, to stop
the baneful torrent.

All this, however, I doubt not, will appear to have
been ultimately for the best. Let temperate and
wise men forwarn the country of its danger, and, as
they are in duty bound, endeavour to prevent, or al-
leviate, evils of every kind. Their conduct will
meet the approbation of the great Governor of the
universe; and, in all events, He, whose will no foreign power can control, being the true and benevolent parent of all the the human race, will provide
for the happiness of his offspring in the most effectual manner, though, to our imperfect understanding,
the steps which lead to it be incomprehensible. We
must not do evil that good may come, because our
understanding is finite, and therefore we cannot be
sure that the good we intend will come. But the
Divine Being, whose foresight is unerring, continually acts upon that maxim, and, as we see, to the
greatest advantage.

To

To the Editor of the Monthly Magazine.

SIR,

I HAVE just read in the Monthly Review, vol. 36, p. 357, that the late Mr. Pennant said of Dr. Franklin, that, " living under the protection of our mild government, he was secretly playing the incendiary, and too successfully inflaming the minds of our fellow-subjects in America, until that great explosion happened, which for ever disunited us from our once happy colonies."

As it is in my power, as far as my testimony will be regarded, to refute this charge, I think it due to our friendship to do it. It is probable that no person now living was better acquainted with Dr. Franklin and his sentiments on all subjects of importance, than myself, for several years before the American war. I think I knew him as well as one man can generally know another. At that time I spent the winters in London, in the family of the Marquis of Landsdown, and few days passed without my seeing more or less of Dr. Franklin; and the last day that he passed in England, having given out that he should depart the day before, we spent together without any interruption, from morning until night.

Now

Now he was so far from wishing for a rupture with the colonies, that he did more than most men would have done to prevent it. His constant advice to his countrymen, he always said, was "to bear every thing from England, however unjust;" saying, that "it could not last long, as they would soon outgrow all their hardships." On this account Dr. Price, who then corresponded with some of the principal persons in America, said, he began to be very unpopular there. He always said, "If there must be a war, it will be a war of ten years, and I shall not live to see the end of it." This I have heard him say many times.

It was at his request, enforced by that of Dr. Fothergil, that I wrote an anonymous pamphlet, calculated to shew the injustice and impolicy of a war with the Colonies, previous to the meeting of a new Parliament. As I then lived at Leeds, he corrected the press himself; and, to a passage in which I lamented the attempt to establish arbitrary power in so large a part of the British Empire, he added the following clause, "To the imminent hazard of our most valuable commerce, and of that national strength, security, and felicity, which depend on union and on liberty."

E e   The

The unity of the British Empire in all its parts was a favourite idea of his. He used to compare it to a beautiful China vase, which, if once broken, could never be put together again : and so great an admirer was he at that time of the British Constitution, that he said he saw no inconvenience from its being extended over a great part of the globe. With these sentiments he left England ; but when, on his arrival in America, he found the war begun, and that there was no receding, no man entered more warmly into the interests of what he then considered as *his country*, in opposition to that of Great Britain. Three of his letters to me, one written immediately on his landing, and published in the collection of his *Miscellaneous Works*, p. 365, 552, and 555, will prove this.

By many persons Dr. Franklin is considered as having been a cold-hearted man, so callous to every feeling of humanity, that the prospect of all the horrors of a civil war could not affect him. This was far from being the case. A great part of the day above mentioned that we spent together, he was looking over a number of American newspapers, directing me what to extract from them for the English

lish ones; and, in reading them, he was frequently
not able to proceed for the tears literally running
down his cheeks.   To strangers he was cold and reserved; but where he was intimate, no man indulged more in pleasantry and good-humour.   By this
he was the delight of a club, to which he alludes in
one of the letters above referred to, called the *Whig-
Club*, that met at the London Coffee-house, of which
Dr. Price, Dr. Kippis, Mr. John Lee, and others
of the same stamp, were members.

Hoping that this vindication of Dr. Franklin will
give pleasure to many of your readers, I shall proceed
to relate some particulars relating to his behaviour,
when Lord Loughborough, then Mr. Wedderburn,
pronounced his violent invective against him at the
Privy Council, on his presenting the complaints of
the Province of Massachusetts (I think it was) a-
gainst their governor.   Some of the particulars may
be thought amusing.

On the morning of the day on which the cause
was to be heard, I met Mr. Burke in Parliament-
street, accompanied by Dr. Douglas, afterwards
Bishop of Carlisle; and after introducing us to each
other, as men of letters, he asked me whither I was

E e 2   going

going; I said, I could tell him whither I *wished to go.* He then asking me where that was, I said to the Privy Council, but that I was afraid I could not get admission. He then desired me to go along with him. Accordingly I did; but when we got to the anti-room, we found it quite filled with persons as desirous of getting admission as ourselves. Seeing this, I said, we should never get through the crowd. He said, " Give me your arm ;" and, locking it fast in his, he soon made his way to the door of the Privy Council. I then said, Mr. Burke, you are an excellent leader; he replied, " I wish other persons thought so too."

After waiting a short time, the door of the Privy Council opened, and we entered the first; when Mr. Burke took his stand behind the first chair next to the President, and I behind that the next to his. When the business was opened, it was sufficiently evident, from the speech of Mr. Wedderburn, who was Counsel for the Governor, that the real object of the Court was to insult Dr. Franklin. All this time he stood in a corner of the room, not far from me, without the least apparent emotion.

Mr. Dunning who was the leading Counsel on the

part

part of the Colony, was so hoarse that he could
hardly make himself heard ; and Mr. Lee, who was
the second, spoke but feebly in reply ; so that Mr.
Wedderburn had a complete triumph.—At the sal-
lies of his sarcastic wit, all the members of the Council, the President himself (Lord Gower) not excepted, frequently laughed outright.   No person belonging to the Council behaved with decent gravity, except Lord North, who, coming late, took his stand
behind the chair opposite to me.

When the business was over, Dr. Franklin, in
going out, took me by the hand in a manner that in-
dicated some feeling.   I soon followed him, and,
going through the anti-room, saw Mr. Wedderburn there surrounded with a circle of his friends
and admirers.   Being known to him, he stepped
forward as if to speak to me ; but I turned aside,
and made what haste I could out of the place.

The next morning I breakfasted with the Doctor,
when he said, "He had never before been so sensible of the power of a good conscience ; for that if
he had not considered the thing for which he had
been so much insulted, as one of the best actions of
his life, and what he should certainly do again in the

same circumstances, he could not have supported it; He was accused of clandestinely procuring certain letters, containing complaints against the Governor, and sending them to America, with a view to excite their animosity against him, and thus to embroil the two countries. But he assured me, that he did not even know that such letters existed, until they were brought to him as agent for the Colony, in order to be sent to his constituents; and the cover of letters, on which the direction had been written, being lost, he only guessed at the person to whom they were addressed by the contents.

That Dr. Franklin, notwithstanding he did not shew it at the time, was much impressed by the business of the Privy Council, appeared from this circumstance :—When he attended there, he was dressed in a suit of Manchester velvet; and Silas Dean told me, that, when they met at Paris to sign the treaty between France and America, he purposely put on that suit.

Hoping that this communication will be of some service to the memory of Dr. Franklin, and gratify his friends, I am Sir, your's &c.

J. PRIESTLEY.

Northumberland, Nov. 10*th,* 1802.

HAVING in my defence of Dr. Franklin, published in the Monthly Magazine, for February 1803, mentioned a circumstance which implied that at that time there subsisted a considerable degree of intimacy between me and *Mr. Burke;* and several persons will wish to know how that intimacy came to terminate, and what could be the cause of the inveteracy with which some years before his death he took every opportunity of treating me, especially by studiously introducing my name, in a manner calculated to excite the strongest resentment, in his speeches in the House of Commons, to which he knew it was not in my power to make any reply, I have no objection to giving the best account that I can of it. It shall be distinct, fair, and candid.

We were first introduced to each other by our common friend Mr. John Lee, while I lived at Leeds, and we had then no difference of opinion whatever, that I knew of, on any subject of *politics*, except that he thought the power of the crown would be checked in the best manner by increasing the influence of the great whig families in the country; while I was of opinion that the same end which we both aimed at would be most effectually secured by a more equal

repre-

representation of the Commons in Parliament. But this subject was never the occasion of any discussion, or debate, between us, except at one time, in the presence of Mr. Lee, at Mr. Burke's table; and this was occasioned by a recent publication of his, on the cause of the *discontents* which then prevailed very generally in the kingdom; a pamphlet of which neither Mr. Lee nor myself concealed our disapprobation, thinking the principles of it much too aristocratical.

When the American war broke out, this difference of opinion did not seem to be thought of by either of us. We had but one opinion, and one wish, on *that* subject; and this was the same with all who were classed by us among the friends of the liberty of England. On the probable approach of that war, but a few years before it actually took place, being still at Leeds, I wrote two anonymous pamphlets, one entitled *The present state of liberty in Great Britain and the colonies,* which gained me the friendship of Sir George Saville, the good opinion of the Marquis of Rockingham, and many other persons, then in opposition to the ministry. Cheap editions were soon printed of that pamphlet, and they

were

were distributed in great number through the kingdom. Soon after this, at the earnest and joint request of Dr. Franklin and Dr. Fothergil I wrote another pamphlet, entitled, *an Address to Dissenters* on the same subject, one sentence of which was written by Dr. Franklin, who corrected the press, as was mentioned in my last. This pamphlet was circulated with more assiduity, and was thought to have had more effect, than any thing that was addressed to the public at the time. Dr. Franklin said it was his serious opinion, that it was one principal reason with the ministry of that day for dissolving the parliament a year sooner than usual; and at the next meeting of parliament, I heard Lord Suffolk, then secretary of state avow that it was done to prevent the minds of the people from being poisoned by artful and dangerous publications, or some expressions of an equivalent nature.

So far Mr. Burke and I proceeded with perfect harmony, until after I had left the Marquis of Lansdowne and while I was in his family I was careful not to publish any political pamphlet, or paragraph whatever, lest it should be thought that I did it at his instigation, whereas politics was expressly excluded

from

from our connexion. But I thought it right never to conceal my sentiments with respect to events
that interested every body; and they were always in perfect concurrence with those of Mr. Burke,
with whom I had frequent interviews.

The last of these was when I lived at Birmingham;
when being accompanied by his son, he called and
spent a great part of the afternoon with me.

After much general conversation, he took me aside
to a small terrace in the garden in which the house
stood, to tell me that Lord Shelburne, who was then
prime minister, finding his influence diminished, and
of course his situation uncertain, had made proposals
to join Lord North.   Having had a better opportunity of knowing the principles, and character of his
Lordship than Mr. Burke, I seemed (as he must
have thought) a little incredulous on the subject.
But before I could make any reply, he said, " I see
" you do not believe me, but you may depend upon
" it he has made overtures to him, and in writing,"
and without any reply, I believe, on my part (for I
did not give much credit to the information) we returned to the rest of the company.   However, it was
not much more than a month, or six weeks, after
                                                      this

this before he himself did the very thing that, whether right or wrong, expedient or inexpedient (for there were various opinions on the subject) he at that time mentioned as a thing so atrocious, as hardly to be credible.

After this our intimacy ceased ; and I saw nothing of him except by accident.   But his particular animosity was excited by my answer to his book on the *French Revolution*, in which, though he introduced a compliment to me, it was accompanied with sufficient asperity.   The whole of the paragraph, which related to the friends of the revolution in general, is as follows.

" Some of them are so heated with their particular
" religious theories, that they give more than hints
" that the fall of the civil powers, with all the dread-
" ful consequences of that fall, provided they might
" be of service to their theories, would not be unac-
" ceptable to them, or very remote from their wishes.
" A man amongst them of great authority, and cer-
" tainly of great talents, speaking of a supposed alli-
" ance between Church and State, says *perhaps we*
" *must wait for the fall of the civil powers before*
" *this most unnatural alliance be broken. Calami-*
" *tous,*

" *tous, no doubt will that time be. But what con-*
" *vulsions in the political world ought to be a subject*
" *of lamentation, if it be attended with so desirable*
" *an effect?* You see with what a steady eye these
" gentlemen are prepared to view the greatest cala-
" mities which can befal their country ! "

The sentiment, however, of this offensive paragraph with which I closed my *History of the Corrup-*
tions of Christianity, and which has been quoted by
many others, in order to render me obnoxious to the
English government, had no particular, or principal,
view to England ; but to all those countries in which
the unnatural alliance between Church and State
subsisted, and especially those European States
which had been parts of the Roman Empire, but were
then in communion with the Church of Rome. Besides that the interpretation of prophecy ought to be
free to all, it is the opinion I believe of every commentator, that those States are doomed to destruction. Dr. Hartley, a man never suspected of sedition,
has expressed himself more strongly on this subject
than I have done. Nothing, however, that any of us
have advanced on the subject implies the least degree of *ill-will* to any of those countries ; for though

we

we cannot but foresee the approaching calamity, we *lament* it; and, as we sufficiently intimated, that timely reformation would prevent it, we ought to be thanked for our faithful, though unwelcome, warnings.

Though, in my answer to Mr. Burke, I did not spare his *principles*, I preserved all the respect that was due to an *old friend*, as the letters which I addressed to him will shew. From this time, however, without any further provocation, instigated, I believe, by the bigotted clergy, he not only never omitted, but evidently sought, and took every advantage that he could, of opportunities to add to the odium under which I lay.

Among other things he asserted in one of his speeches, that " *I was made a citizen of France on* " *account of my declared hostility to the constitution of* " *England;*" a charge for which there was no foundation, and of which it was not in his power to produce any proof. In the public papers, therefore, which was all the resource I had at that time, I denied the charge, and called upon him for the proof of what he had advanced; at the same time sending him the newspaper in which this was contained, but

he

he made no reply. In my preface to a Fast Sermon in the year following, I therefore said that it sufficiently appeared that " he had neither ability to main- " tain his charge, nor virtue to retract it." This also was conveyed to him. Another year having elapsed, without his making any reply, I added, in the preface to another Sermon, after repeating what I had said before, " a year more of silence on his " part having now elapsed, this is become more " evident than before." This also he bore in silence.

A circumstance that shows peculiar malignity was, that on the breaking out of the *riots of Birmingham* a person who at that time lodged in the same house with him at Margate, informed me that he could not contain his joy on the occasion ; but that running from place to place, he expressed it in the most unequivocal manner.

After this I never heard any thing concerning Mr. Burke, but from his publications, except that I had a pretty early and authentic account of his *first pension*, which he had taken some pains to conceal. Such is sometimes the fate of the most promising, and long continued, of human friendships. But

But if I have been disappointed in some of them, I have derived abundant satisfaction, and advantage from others.

J. PRIESTLEY.

Northumberland, February 1, 1804.

APPENDIX, NO. 5.

A Summary of Dr. Priestley's Religious Opinions.

IT has already been mentioned that it was once the intention of Dr. Priestley to draw up a brief statement of his Theological opinions; not only to prevent misconception and misrepresentation, of which in his case there has been more than a common portion, but also to exhibit a system of Faith, plain, rational and consistent, such as common sense would not revolt at, and philosophy might adopt without disgrace.

This merit at least, (no common one) Dr. Priestley is fairly entitled to in relation to the tenets he ultimately adopted. The prejudices of his youth were to be surmounted in the first instance. He had to encounter, not these only, but the prejudices of his early and most valued connections. Every change of his opinion, was at the time, in manifest hostility with his interest; and every public avowal on his part of what he deemed genuine Christianity, put in jeopardy the attachment of his friends, the

F f

support

support of his family, and his public reputation : nor
was this all : for it subjected him with fearful certainty, to the hue and cry which bigotry never fails to
raise against those who in their search after truth,
are hardy enough to set antient errors, and established absurdity at defiance.

The writings of Dr. Priestley however enable his
readers to do that, which it is much to be regretted
he did not find opportunity to do for himself. Not
that any thing I can now venture to state on this subject will compensate for the accuracy and superior
authenticity of such a confession of faith as he would
have penned for himself; and still less for the ener-
getic simplicity which would have characterized
such a solemn condensation of the researches of half
a century on the most important objects of human
enquiry. But it is not improbable that a fair and
impartial exhibition of the principal points of his religious creed, may serve to shew, that christianity,
such as he conceived it to be in its original purity,
and such as he professed and practised, has a direct
tendency to make men wiser and better, more patient, resigned, and happy here, and affords hopes and
prospects of futurity more cheering than those who
ire not christians can possibly enjoy.

                                                    That

That there are difficulties attending the evidences of christianity, which may give rise to important doubts in the breast even of serious and candid enquirers, no person who has duly studied the subject, and who is not paid for professing the fashionable side of the question, will be hardy enough to deny. Good and wise men have enlisted on both sides of the argument; nor is it an impeachment either of good sense or integrity, to adopt either side. The christian sneerers at honest infidelity, and the philosophic sneerers at rational christianity, appear to me equally distant from that frame and temper of mind which characterizes the real votary of truth.

I shall state then what appears to me, a fair summary of Dr. Priestley's religious creed, premising, that my own assent does not accompany all the tenets which on the maturest investigation, and on the fullest conviction, he adopted as the clearest and most important of truths. I would it were otherwise: but assent is not in our power to give or to withhold. Theology was a subject on which we had agreed to differ: a difference, which though a mutual source of regret, was to neither of us a cause of offence.

Dr. Priestley believed in the existence of one God:

F f 2

one

one Supreme Creator and governor of the universe: eternal in duration; infinite in goodness, in wisdom and in power: to whom, and to whom alone, all honour is due for the good enjoyed by his creatures: to whom, and to whom alone all thanks were to be given for benefits received, and all prayer directed for benefits desired.

He believed, that the system of the universe formed by this being, was the best upon the whole that could have been devised by infinite goodness and wisdom, and executed by infinite power. That the end of creation, in all probability, was the happiness of the sentient and intelligent beings created. That the moral and physical evil observeable in the system, according to men's limited view of it, are necessary parts of the great plan; all tending ultimately to produce the greatest sum of happiness upon the whole, not only with respect to the system in general, but to each individual according to the station he occupies in it. So that, all things, in the language of Scripture, under the superintending providence of the Almighty "work together for good."—A system thus pre-ordained in all its parts, and under the influence and operation of general laws, implies the necessary depend-

dependence of every action and event on some other preceding as its cause until we arrive at the deity himself the first, the great and efficient cause of all. Such a system excludes also, the necessity of particular interpositions of Providence, other than such as might have been foreseen and pre-ordained in the beginning, and embraced within the general plan.

It was the death and resurrection of Christ alone that brought life and immortality to light. On the doctrines of christianity, and on them alone in his opinion a christian must rely for satisfactory evidence of a future state. But independent of the christian Scripture, it resulted from the metaphysical part of his Theology, and he thought it *probable* from the light of nature, that the present life is but an introduction to future states of eternal existence which man is destined to pass through; wherein, virtuous and benevolent dispositions and increased knowledge, will constitute the means of conferring and of enjoying happiness; and that evil, of whatever kind, is permitted to exist among intelligent beings, because necessary as a means of eradicating vicious propensities, and of gradually introducing in their stead those habits of virtue and benevolence, without which

F f 3

happi-

happiness cannot exist either here or hereafter. The future happiness of individuals, will therefore depend on the degree to which they have cultivated those dispositions here; and the evil they will suffer, will necessarily be in proportion to the vicious habits they have acquired during their passage through the present life. But although he was of opinion that these ideas might now be rendered probable independent of the Scriptures, he was firmly persuaded that the light of nature alone would never have suggested them; since in fact, nothing can be more crude, more unsettled, and unsatisfactory than the notions advanced by the wisest of the heathen philo sophers who had no light to guide their researches, but what is called the light of natnre. It is christianity alone therefore that has suggested those liberal notions of the being and attributes of God and the benevolent plans of divine providence, which we are grossly mistaken if we conclude the light of nature would have pointed out, though it may serve in some degree to strengthen and confirm.

It has been necessary (as he thought) to the present and future welfare of the human race, and a part of the system ordained by the Almighty from the beginning

ginning that in consequence of the prevalence of ignorance and vice in the world, teachers, preternatu-
rally endowed with wisdom and power by God himself, should occasionally appear for the purpose of
promulgating more correct notions of the being and
attributes of the Almighty, and of the duties of men
toward their maker and toward each other. In particular, to assert the unity of the Divine Being in opposition to the idolatrous worship and polytheistic notions of the pagan world, and to furnish a more sure
and compleat sanction to morality.*

That these preternatural interpositions in favour
of the human race, were more especially manifested
in the Jewish dispensation by means of Moses, and
in the christian dispensation by means of Jesus
Christ: both of whom were especially commissioned
by God for the purpose; and each of these dispensations being respectively calculated for the state and
condition

---

* I was for a long time satisfied with Warburton's Hypothesis, that
under the Jewish Theocracy, there was no sanction but that of *temporal* reward and punishment. I do not recollect Jortin's or Sykes'
pamphlet in reply. But a small tract written by Dr. Priestley on
this subject, one of his latest compositions, convinces me that Warburton's opinion was very probably, if not demonstrably erroneous.

condition of mankind, at the time when these holy men appeared. That profane history, abundantly proves this necessity ; and the utter inability of human wisdom in its best state at the time, to arrive at those correct ideas of religion and morality, which it was the end of those dispensations to communicate and sanction.

That the books of the old and new Testament contain the history of those dispensations, and the circumstances attending them, so far as it is necessary for us to be made acquainted with the facts. These books are the histories of the times treated of, by various writers: written from the common motives which have dictated other histories: without any pretence to divine inspiration (except in the case of the prophecies) : and are to be tested by the same canons of criticism, by which we determine in other cases, whether a book be really written by the author to whom it is ascribed, and whether the material facts related, are accompanied with sufficient evidence internal and external, to justify our giving credit to them. He believed, that these books like other histories though far less antient, may have suffered in many passages of small moment by frequent transcription

cription and unauthorized interpolation: that the
authors, like other observers, might commit mistakes and differ from each other in particulars of minor import; but there is evidence as strong, nay
much stronger both internal and external of the accuracy and fidelity of the writers, and of the truth of
the material facts related in these books, than in the
case of any other history extant, judging by the common rules that an unbeliever would adopt for decid-
ing the question. Considering the subject in this
way, he did not adopt as canonical every passage
indiscriminately contained in the old and new Testament, but rejected some that were not accompanied
with equal evidence of authenticity with the rest.
Hence he did not believe in the history of the miraculous conception; or in the interpolated passage of
the three witnesses; nor indeed could he embrace
the polytheistic doctrine of the Trinity in any shape,
when he regarded the absolute Unity of God, as the
great doctrine, the characteristic feature of revealed
religion, uniformly taught by Moses and the Prophets, as well as by Christ and the Apostles, in opposition to the polytheism of the Pagans.*

---

* He admitted the " Revelations" into his list of canonical Books;
though I do not think he was sufficiently aware of the objections of
Abauzit.

From a careful comparison of Scripture with itself he deduced his opinion that Christ like Moses was a mere Man, divinely commissioned to preach and propagate a sublimer religion, and a purer morality than had yet been known: and for the purpose of giving force and effect to the doctrines he taught, power also was given to him to perform in the eyesight, and under the observation of multitudes opposed to his pretensions, actions of kindness and benevolence toward individuals, that no human means could accomplish. All which were abundantly confirmed as proofs of his being sent of God by his foretelling his own death and resurrection in the time and manner as they actually took place.

Thus far he believed the mass of testimony fully bore him out in giving his assent to the divine mission of Jesus Christ, and to the doctrines he taught. A mass of testimony which if false or forged, constituted in his opinion (judging from the common principles of human nature, and the acknowledged rules of evidence) a miracle far more incredible than any that christianity requires to be believed. He saw no reason however for believing that either Moses or Jesus Christ were inspired with supernatural know-

knowledge, or endued with supernatural power, beyond the immediate objects of their mission. When the reason and the occasion ceased, the supernatural gifts would cease too. They were given for a certain purpose : we are not warranted therefore in extending them beyond the occasion that called them forth.

In the same manner he thought of the Apostles, notwithstanding the high authority that accompanied their opinions, from their situation of intimacy with Jesus Christ. Yet when reasoning from themselves and as men, they would sometimes like other men be liable to reason inconclusively. That they did so sometimes must be allowed from the manifest differences of opinion among each other on some of the less important points of christian practice and doctrine.

In examining the language of scripture, he made due allowances as a man of learning and good sense ought to do, for the peculiar idioms, allusions, and figures, which though not likely to mislead or be misunderstood by the persons to whom they were addressed, will not now bear a literal interpretation consistent with the known attributes of the Supreme

Being,

Being, and the immutable principles of right and wrong. Hence he rejected the gloomy doctrine of Original Sin, as well as the strange hypothesis of vicarious suffering, or the doctrine of Atonement. No system of religion however apparently founded on miraculous evidence, can require us to believe, that the axioms of moral justice, any more than of the mathematics can be false. It would seem as difficult to demonstrate that one man ought to be punished for the offences of another with whom he has no connexion, as that a part was equal to the whole, or that two quantities each equal to a third were unequal to each other. His accurate search into biblical phraseology, fully satisfied him that these strange tenets of what is called Orthodoxy, were equally unfounded in scripture and common sense.

For the same reason he rejected the horrid criterion of Calvinistic Theology, the doctrine of election and reprobation, and its concomitant, the eternal duration of future punishment. Indeed, he had no notion of punishment as such in the common acceptation of the term. The design of the Creator in his opinion, was the ultimate happiness of all his creatures by the means best fitted to produce it. If

pain

pain and misery be the consequence of Vice, here or hereafter, it is nevertheless an instance of God's fatherly kindness toward the creature who suffers it, because that suffering is absolutely necessary to eradicate the dispositions that obstruct the progress of improvement in knowledge and virtue, and close all the avenues to real happiness. Punishment therefore, is not inflicted with the slightest tincture of revenge, but as a necessary means of qualifying the sinner for a better state of existence, which his present propensities disqualify him from enjoying. It is not the effect of anger in an irritated and avenging tyrant as the abominable tenets of Orthodoxy would induce us to think of the Deity, but it is the *medecina mentis* exhibited for our good by the Physician of Souls. Nor have we any reason to believe that it is greater in degree, or longer in duration, than is necessary to produce the beneficial effect for which it is inflicted. It is that sort of punishment which a kind but wise parent, inflicts on a beloved child.

At one time indeed, he seems to have entertained the opinion that annihilation might possibly be the lot of the wicked: but deeper reflection, and the fair results deducible from his metaphysical as well as his

his theological system, altered his opinion. Trusting therefore to that pre-eminent and delightful attribute of the Deity—that attribute to which wisdom and power are but the handmaids, the Divine BENEVOLENCE, he did not doubt but the ultimate result of the system would be permanent happiness to every intelligent being it embraces, though through different trials, at different periods and perhaps in different degrees. This doctrine he found as conformable to the scriptures as it is to just notions of the goodness of God ; and it seems to furnish a glorious exposition of that cheering passage, GOD IS LOVE;

Thus persuaded, that happiness essentially consists in conferring happiness, and that our only notion of any source of happiness to the Deity is the infinite power he possesses of communicating it to his creatures, no wonder he was impressed himself, and endeavoured to impress others with the *Duty of having God in all our thoughts*, and, *The duty of not living to ourselves* : sentiments illustrated with a degree of energy and conviction never exceeded, in two of the finest sermons ever composed, and to which he gave these titles. It was this that animated him to inces-

sant

sant exertion in the pursuit and the communication
of knowledge of every kind : for knowledge he considered as equivalent to power, and as the most extensive and effectual means of doing good to others,
certainly here, and probably hereafter.

These were the doctrines that he adopted and
taught; doctrines, not merely professed, but deeply
felt, and daily acted upon. This it was, that taught
him habitually to regard every event as ultimately a
blessing; that drew the sting of misfortune, and al-
layed the pang of disease. He felt indeed for a time
as others feel in similar circumstances; but his
mind soon recovered its tone, and applied with salu-
tary effect to the ideas so long cherished, and so inde-
libly impressed, that God orders all things for good.
This was a consolation to which he never resorted in
vain.

These seem to me the most important and prominent features of the system he professed, nor is it
worth while to dwell upon the minuter points in
which he differed either from the established church
or the Dissenters. In Church Government he was
an Independent, believing that any number of pious
christians meeting together for the purposes of public

lic worship formed a Church, *Cœtus credentium;* of which the internal regulation belonged to the persons composing it.   He never I believe, either prayed or preached extempore; conceiving every Pastor at liberty in this respect to follow that practice which he found most tending to edification.   He was a friend to infant Baptism, and to exhibiting the commemoration of the Lord's supper to young people, for reasons assigned in the pamphlets he published on these subjects.   He not only believed the keeping of the Sabbath to be a duty incumbent on christians, and having in its favour the practice of the earliest professors of christianity, but he was a strenuous advocate for family prayer, which he constantly attended to in his own family.

His opinions respecting the soul, of course led him to disbelieve the doctrine of an intermediate state. Believing that as the whole man died, so the whole man would be called again to life at the appointed period of the resurrection of all men, he regarded the intermediate portion of time as a state of utter insensibility : as a deep sleep, from which the man would awaken when called on by the Almighty, with the same associations as he had when alive, without being

ing sensible of the portion of time elapsed. He did not think the light of nature sufficient to furnish satisfactory evidence of any future state of existence, and therefore the christian scriptures which alone gave full conviction, and certainty on this most important point, were to him peculiarly and proportionably dear. To him, a future state was a subject of ardent and joyful hope, though to the majority of those who believe and contemplate the gloomy doctrines of orthodox christianity, it cannot but be a subject of frequent and anxious dread, and of very dubious and uncertain desire.

Such were the chief of Dr. Priestley's tenets on the subject of Religion. Be they true or false, they were to him a source of hope and comfort and consolation : his temper was better, his exertions were greater, and his days were happier for believing them. The whole tenor of his life was a proof of this ; and he died resigned and cheerful, in peace with himself and with the world, and in full persuasion that he was about to remove to as phere of higher enjoyment, because it would furnish more extensive means of doing good.

G g

ERRATA.

Page	Line				
23	10 from the top,	For deliverery,	read delivery.
84	14 — —	Actes,	Artes.
	— —	pecsinit,	nec sinit.
90	bottom line,	No. 6,	No. 4.
160	2 — —	Bur,	But.
172	3 — —	Liancount,	Liancourt.
188	1 — —	determing,	determining.
	8 — —	he,	be
214	1 — — For wall,	read well.
218	3 from the bottom,	immorality,	immortality.
329	2 — —	1679,	1767.
289	3 — —	fort,	forte.
304	8 from the bottom after, the Author,	Dr. Coward.
	10 — —	Philabethes	Psycalethes.
	1 from the top	predomninates	predominates.
333	7 from the top of the note for dise,	dire.
357	12 from the top	For is,	read it.
422	2 from the top	*confervu*	*conferva.*
	3 from the bottom	Hydertids,	Hydatids.
423	from the top	nydra,	hydra.
425	9 from the bottom	for Ἰουλα	Ἰουλα
426	4 from the top	terms,	terrors.
	14 from the bottom	acccrate,	accurate.
444	5 from the top	flower,	slower.

Printed in the United States
By Bookmasters